T0002897

TOMORROW'S LAWYERS

TOMORROW'S LAWYERS
An Introduction to Your Future

THIRD EDITION

Richard Susskind

OXFORD
UNIVERSITY PRESS

OXFORD
UNIVERSITY PRESS

Great Clarendon Street, Oxford, OX2 6DP,
United Kingdom

Oxford University Press is a department of the University of Oxford.
It furthers the University's objective of excellence in research, scholarship,
and education by publishing worldwide. Oxford is a registered trade mark of
Oxford University Press in the UK and in certain other countries

First Edition published in 2013
Second Edition published in 2017
Third Edition published in 2023

Impression: 1

Published in the United States of America by Oxford University Press
198 Madison Avenue, New York, NY 10016, United States of America

British Library Cataloguing in Publication Data
Data available

Library of Congress Control Number is on file at the Library of Congress

ISBN 978-0-19-286472-7

Printed and bound in the UK by
Clays Ltd, Elcograf S.p.A.

I dedicate this book to
Daniel, Jamie, and Ali,
my loving children,
who bring me endless happiness

PREFACE TO
THE THIRD EDITION

Rather disconcertingly, I received some congratulatory emails during the pandemic. These messages noted that lawyers had become big users of technology because of COVID, and they insisted in consequence that, 'Richard, your future has arrived'. By implication, they also seemed to be suggesting that it might be time for me to hang up my hat.

I disagree with both propositions.

I do not think that lawyers have arrived at their final destination simply because many now regularly work from home. Mastering video calls and working remotely are but early steps towards the transformed legal landscape that I envisage in my books. Nor do I think—I am relieved to report—that my work is done. Indeed, I extend my heartfelt thanks to the many lawyers and law teachers around the world who have encouraged me to prepare this third edition.

Will there ever be a day, though, when tomorrow's lawyers will be today's lawyers, and so no further editions will be required? To ask that question is to misunderstand the era in which we now live and labour. Over and above changes that are brought by the horrors of pandemic and war, technology is advancing at such a ferocious rate that, for the foreseeable future, we must expect to be operating in law in a period of endless flux. As our systems become increasingly capable, some current legal jobs will fade, new jobs will doubtless emerge, although many of these, in due course, will also be taken on by machines. From here on in, lawyers will need continually to flex and adapt to keep in demand and remain relevant.

As to the speed of change, that is much trickier to predict. I completed the first edition of this book a decade ago. This means that we are now half-way through the 20-year period, during which, I originally predicted, legal institutions and lawyers would change more than they had in the past two centuries. I never thought that this was such an extreme claim because, in truth, the world of law has not changed much in the last 200 years, or at least, in England, since the reforms of the 1870s. In any event, I see no need to revise my forecast. As discussed throughout the book, looking across the entire legal ecosystem, we have already seen significant changes, and the pace of change—and this is vital—continues to accelerate. One modest bit of evidence to support this claim: I had to work much harder to revise the book this time round. A decade on, there were many more changes to note and make than after five years.

That said, the world's outstanding law firms are still evolving at a sedate pace. I know that many of their leaders say that they have changed their businesses substantially over the last few years. But I believe what we have seen recently in these (massively successful) firms have been some operational shifts at the edges, and some good marketing too, rather than full-scale transformation. However, the early shoots of change are now visible and I anticipate that the 2020s will witness great shifts in the business models underpinning these major firms.

I expect the trajectory of change in law to follow an exponential curve. Over the past decade, we have seen a shallow, steady change and are now seeing signs of more explosive movement. We are at the knee of the curve.

Looking longer term, what would count for me as radical change and transformation in law firms? In answer to that question, let me propose a test, according to which a law firm has radically changed when less than half of its revenue is derived from charging for the time of traditional work of human lawyers. The fundamental revolution in law (and other professions) will come about when we move from one-to-one consultative advisory service to one-to-many online or embedded service. Income will come not from selling people's time but from licensing content. No law firm in the world currently comes close to passing my proposed test. It may well be that no firms will ever change to this extent. But if they do not, I am confident that other alternative legal providers will instead grasp the opportunity—to serve their communities better and, as I say in the body of the text, to make money while they sleep.

Finally, I should say that my own thinking has advanced since 2017, when the second edition of this book first appeared. I have since written a new work, entitled *Online Courts and the Future of Justice* (2019 and 2021), and, with Daniel Susskind, have published an updated edition of *The Future of the Professions* (2022). Some of the ideas and findings of these books have found their way into this edition.

Although I try to keep the book short and punchy, I find myself straying further from this aspiration with each edition. No matter its length, though, I stress that this book is introductory—conceived as a brisk canter through the world of tomorrow's lawyers. Others are now doing a great job of putting more flesh on the bones. My purpose, as always, is to encourage conversation and then action.

Richard Susskind
August 2022
Radlett, England

PREFACE TO THE SECOND EDITION

One of the central claims in the first edition of this book, published in 2013, was that the legal world would change more in the next 20 years than it has in the past two centuries. Three years on, I believe we are on course. In the intervening period, much has happened in the legal world. To give a flavour—many major law firms have since set up low-cost service centres to undertake routine legal work; the Big 4 accounting firms have rapidly grown their global legal capabilities; there has been a great upsurge of legaltech start-ups, now well over 1,000 worldwide; the idea of artificial intelligence (AI) in law has captured the imagination of innovators across the profession, from market leading firms to law student developers; in England and Wales, in our liberalized legal regime, innumerable 'alternative business structures' have been launched (there are now over 500); professional bodies, such as the Canadian Bar Association, have produced studies on the future of legal services; senior judges have been strongly advocating the wider use of technology; the British government has committed to investing over £1 billion in modernizing and digitizing the court system in England and Wales; innumerable in-house legal departments, especially in the US, have been appointing chief operating officers to rethink and manage their operations; and, if readers will forgive me, more Chinese lawyers have bought this book than English lawyers. In short, much has changed in these three short years. At the same time, most commentators agree that the pace of change is accelerating. And great numbers of leaders across the legal profession are now openly acknowledging that the world of law is entering a period of transformation.

In truth, we are just warming up.

My own thinking has also moved on. In part, I have learned, as always, from my law firm clients, each of whom has been powering ahead, embracing alternative labour models as well as new technologies. I have also been preoccupied over the last couple of years with online dispute resolution. I chaired an advisory group of the Civil Justice Council that devised the idea of online courts, and their introduction is now judicial and government policy for England and Wales. Above all, I had the great privilege of writing a book, *The Future of the Professions*, with my son, Daniel Susskind, a Fellow in Economics at Balliol College, Oxford. This was published in 2015. Looking beyond law to other professions, and working with an economist, has led me to revisit some of my past analysis.

All of which is to say that the first edition of this book is now out of date. Accordingly, my aim in writing this second edition is to take account of recent advances in the legal marketplace alongside developments in my own thinking and experience. Once again, I have kept it brief, because my concern here is with the big picture—the broad trends and the likely outcomes. My main purpose is to encourage open-minded debate and reflection, with a view to improving our legal systems. While the book was originally intended for young lawyers, it transpires that older lawyers also like short and inexpensive books and so they read it too. I am happy with this—all lawyers, unless they are retiring today, are tomorrow's lawyers.

Richard Susskind
October 2016
Radlett, England

PREFACE TO
THE FIRST EDITION

I have written this book to provide tomorrow's lawyers and legal educators with an accessible account of the pressing issues that currently face the legal profession and the justice system. We are, I have no doubt, on the brink of fundamental change in the world of law, and my main aim is to encourage wider discussion of the forces at play and their likely impact.

Although originally conceived as a guide to the future for the next generation of lawyers, I expect that the book will also be read by more experienced practitioners. For busy professionals who do not have the time to read lengthy texts, I hope that it serves as a punchier version of my ideas than my previous work. Certainly, it represents a substantially updated version of my views on trends in the legal market.

I do not anticipate that readers will agree with all of what I say. But if the book gives rise to more serious reflection and debate about the future of the law and lawyers, then I have done my job. And yet, because we live in such rapidly shifting times, it is a job that is necessarily incomplete. Each day, I hear fresh tales of innovation in law—a new legal business here, an online facility there, and a regular flow of imaginative ideas for meeting clients' needs in different ways. In citing these innovations, I had to draw the line somewhere, however, and so I have only been able to refer to developments that surfaced before the end of May 2012. I would not be surprised, by the time this book is published, if some important new legal services have been launched in the interim.

I have some people to thank. First of all, there is the team at Oxford University Press. This is the fifth time that OUP has agreed to take on one of my books and, as ever, it has been a privilege to work with such a well-regarded publishing house. I am especially grateful to Ruth Anderson and Sophie Barham in the UK and to Ninell Silberberg in the US for their friendly support and advice. I must also record my thanks to the various referees who anonymously assessed my book proposal and made a wide range of suggestions that led, I believe, to many significant improvements.

Next is Patricia Cato, who helped me with innumerable initial drafts and still comfortably outperforms any speech recognition system in making sense of my rapid Glaswegian.

I have also benefited greatly from the guidance, encouragement, and criticisms of a small group of friends and colleagues who generously spent many hours of their time reading an early draft of the book—Neville Eisenberg, Hazel Genn, Daniel Harris, Laurence Mills, David Morley, Alan Paterson, and Tony Williams. To each, I extend my profound thanks.

Two reviewers deserve separate mention—my sons, Daniel and Jamie. This book would not have been written without their love and encouragement. They enthused when I came up with the idea of a book for aspiring lawyers, they motivated me when other commitments made it difficult to maintain momentum, and they commented extensively on earlier drafts. Their range and clarity of thought amaze me.

The last person but one to thank is Ali, my daughter and friend, to whom, along with her brothers, this book is dedicated. I cherish every one of the many companionable moments we spend together. I could not have a more wonderful daughter.

And finally, as always, I am very grateful to my loving wife, Michelle. For over 30 years now, she has indulgently endured my bouts of obsessive writing. It cannot be easy. Her boundless support for my work and her confidence in my ideas mean so very much to me.

Richard Susskind
June 2012
Radlett, England

When one door closes, another door opens; but we often look so long and so regretfully upon the closed door that we do not see the ones which open for us.

ALEXANDER GRAHAM BELL

Institutions will try to preserve the problem to which they are the solution.

CLAY SHIRKY

If you don't know where you're going, you might not get there.

YOGI BERRA

CONTENTS

Introduction *1*

PART ONE Radical Changes in the Legal Market

PART TWO The New Landscape

PART THREE Prospects for Young Lawyers

| Introduction

This book is a short introduction to the future for young and aspiring lawyers. Tomorrow's legal world, as predicted and described here, will bear little resemblance to that of the past. I stand by my claim of 2013, in the first edition of this work, that legal institutions and lawyers would change more radically in less than two decades than they had in the past two centuries.

If you are a young lawyer, the revolution I discuss is happening on your watch. 'Young' should be construed broadly, applying to students who are contemplating a job in law through to newly promoted partners in firms who are wondering how their careers might unfold. I also write for those who are interested in young legal businesses, such as the lawtech start-ups and the alternative legal providers that are now seeking to redefine the legal marketplace.

To elder statesmen in traditional firms, who may feel after a couple of paragraphs that they are excused from reading on, I issue a warning. Although it may appear that the future, and particularly the topic of technology, is of interest primarily to the next generation, some of the transformations that I discuss here are coming in the next few years if not with us already. Unless retirement is imminent, what I say here will directly affect older lawyers too. More than this, leaders in the legal profession today should be concerned not just about hanging on until their pensions click in, but about their long-term legacy as well.

'My call is to the young in heart, regardless of age', John F. Kennedy once said, and I say this again now. I write primarily for the youthful of spirit, for the energetic, for the optimistic— for those who join me in recognizing that we can and should modernize (I now like to say 'upgrade') our legal and justice systems.

Discontinuity in the Legal Profession

This book comes at a time of great debate in the legal world over an array of vital issues. There is deep concern, for example, about cuts in public legal funding that may reduce 'access to justice'. There are worries about law schools that seem to be offering costly places to students in greater numbers than there are job opportunities. And there is unease about the disproportionate expense of pursuing claims in the courts.

I offer remedies for these and many other ills, but I do not provide the same sorts of answers as those offered by most careers advisers, parents, professors, and legal practitioners. For example, while most lawyers are arguing for smaller cuts in legal aid, I argue we should be exploring and implementing alternative ways of providing legal guidance, not least through online legal services; while commentators agitate about over-recruitment into law schools, I identify a whole set of exciting new occupations for tomorrow's lawyers, although I am troubled that we are not preparing students and young practitioners for these jobs; and while judges and litigators are seeking to control the costs of litigation, I believe we should be increasing our use of virtual hearings, online courts, and online dispute resolution.

Most inhabitants of today's legal world tend to look for solutions by extrapolating from the past and on the assumption of continuity in the legal profession. In contrast, I foresee discontinuity over time and the emergence of a legal industry that will be quite alien to the current legal establishment. The future of legal service is neither Grisham nor Rumpole. Nor is it wigs, wood-panelled courtrooms, leather-bound tomes, or arcane legal jargon. It will not even be the dominant model of lawyering, which is one-to-one, consultative professional service by advisers who dispense tailored counsel. To meet the needs of clients, we will need instead to dispose of much of our current cottage industry and reinvent the way in which legal services are delivered. Just as other professions are undergoing massive upheaval, then the same must now take place in law. Indeed, it is already happening. The bespoke specialist who handcrafts solutions for clients will be challenged by new working methods, characterized by lower labour costs, mass customization, recyclable legal knowledge, pervasive use of advanced technology, and more.

When I was at law school, in the late 1970s and early 1980s, few students gave much thought to what the future might hold for the legal profession. We took it for granted that the work of lawyers in, say, 25 years' time would be much as it was in our time. It transpired that we were right to expect little change. In contrast, in looking 25 years ahead from now, I argue it would be absurd to expect lawyers and courts to carry on operating as they do now. If only because of the inexorable rise in the power and uptake of technology—to pick one of several drivers of change—we must surely expect something manifestly more than modest adjustment.

So Why Listen to Me?

You might think that hordes of senior people in the legal profession are currently thinking deeply about the long-term prospects for lawyers and the legal system. But very few of the people whom you might expect to be at the helm—politicians, senior partners in law firms, policy-makers, law professors, top judges, leaders of professional bodies—are actually looking much further than beyond the next few years. In these difficult times, the here-and-now seems to be providing headache enough.

In truth, in the legal community there are only a few hundred lawyers and professors around the world who are devoting their full working lives to theorizing about and planning for the long term (some of their works are referenced in the Further Reading section of this book). I am one of them and have been writing, speaking, and advising on the future for longer than most. I started my journey in 1981, as a third-year undergraduate law student at the University of Glasgow. Since then, I have written a doctorate in law and computers at Oxford University, and worked for several years with one of the 'Big 4' accounting firms and then for much of the 1990s with an international law firm, on whose board I sat for three years. I have been a law professor for more than three decades, and, for 25 years, an independent adviser to international law firms, national governments, in-house legal departments, and judiciaries around the world. And I have been Technology Adviser to the Lord Chief Justice of England and Wales since 1998.

Even my fiercest critics will concede that in my numerous books and newspaper columns over the last 35 years I have been right more often than wrong in my predictions. So, I say this: if there is a better than even chance that the radically transformed legal world I predict here will come to be, then it should be worth spending a few hours contemplating its implications. If my winning run continues—and my confidence in my predictions is greater now than in the 1990s—then it might pay dividends to read on. And my hope is that readers will not respond defensively ('how can we stop this happening?') but will find exciting new options and opportunities in these pages ('I want to be one of the pioneers').

How the Book is Organized

The book is divided into three main parts. The first is an updated, simplified restatement of my views on the future of legal services, as presented in four other works—*The Future of Law* (1996), *Transforming the Law* (2000), *The End of Lawyers?* (2008), and *The Future of the Professions* (2015, co-authored with Daniel Susskind and updated in 2022). I have tried to pick out and highlight the key themes of these books. I introduce the main drivers of change in the legal market and explain why and how these will lead lawyers to work differently and will encourage new providers to enter the market with novel approaches to legal service. Along the way, I consider the impact of the pandemic on the legal world and then outline a wide range of technologies that I believe will disrupt the traditional working practices of lawyers. To help put these systems in context, I provide a framework for thinking in a structured way about the impact of legal technology. My focus here, as throughout, is largely, but by no means exclusively, on civil work in commercial law firms. For readers who are already familiar with my ideas from my other books, I urge you not to skip Part One, because there have been significant developments in the market and in my thinking since I last wrote about these.

Next, in Part Two, I sketch out the new legal landscape, as I expect it to be. I discuss the future for law firms, the challenges facing in-house lawyers, and the emergence of a burgeoning lawtech start-up industry. In summary form, I suggest some timings for the shifts I anticipate. I also lay out some of the ways in which 'access to justice' problems will be overcome through a variety of online legal services. I offer some predictions too about the work of judges and the courts, and the promise of video hearings, online courts, and online dispute resolution. Here I draw on my most recent new work, *Online Courts and the Future of Justice*, published originally in late 2019 but, such were the changes wrought by the pandemic, updated in 2021. I also take the opportunity to revisit the findings of my book, *The Future of Law* (1996), where I made some 20-year predictions about the legal world.

Lastly, in Part Three of the book, I focus more specifically on the prospects for young lawyers. I ask what new jobs and new employers there will be, and for what and how the next generation of lawyers will be trained. I believe that I provide optimistic and encouraging answers to these questions. I also equip young lawyers with some penetrating questions to put to their current and prospective employers. I then offer some thoughts on 'innovation', a topic on the lips of many lawyers and certainly their marketing advisers. Finally I look to the long term, not least at artificial intelligence (AI), and I put down a challenge for young (in heart) lawyers everywhere.

Wayne Gretzky, perhaps the finest ice hockey player of all time, famously advised to 'skate where the puck's going, not where it's been'. Similarly, when lawyers are thinking about the future, whether about their law firms or law schools, they should be planning for the legal market as it will be and not as it once was. In ice hockey terms, I worry that most lawyers are still skating to where the puck used to be. My purpose, then, is to show where that puck is most likely to end up.

PART ONE

Radical Changes in the Legal Market

1 | Three Drivers of Change

The legal market is in a remarkable state of flux. New ways of delivering legal services are emerging, new providers are becoming firmly established in the market, and the workings of our courts are being transformed. Unless they adapt, many traditional legal businesses will fail. On the other hand, in these fluid times, a whole set of new opportunities are presenting themselves to entrepreneurial and creative young lawyers.

I believe there are three main drivers of this change. I call these the 'more-for-less' challenge, liberalization, and technology. Other commentators may point to different factors, such as diversity and inclusion, shifting demography, climate change, mental health, threat of war, and globalization. I do not deny that such factors are hugely significant but my specific focus here is on the changes that we will see in *the way in which legal services are delivered*; and all my research and advisory work, as well as what I have seen in other professions, lead me to the conviction that, in relation to how legal work is undertaken, my three drivers are the ones to watch for. Let me introduce each in turn.

The 'More-for-Less' Challenge

Clients of lawyers come in many different forms. There are in-house lawyers, who work within large organizations and who spend mightily on legal advice when they have major disputes to resolve or large deals to conclude (annual legal budgets north of US$1 billion are not unusual). There are managers within small or medium-sized businesses, who have properties to rent, employees to engage, and all manner of regulations with which to comply. And there are individual citizens, who may need legal help with matters such as moving house, coping with debt, or pursuing some personal injury claim. Although diverse in nature, these clients currently share a big challenge—generally, they cannot afford legal services when delivered in the traditional way.

General Counsel, the individuals who run in-house legal departments, invariably say that they face three problems. First of all, because of difficult economic conditions, they are under pressure to reduce the number of lawyers in their teams. Second, they are being asked by their chief executives, chief finance officers, and boards to reduce the amount they spend on external law firms. And yet, at the same time, third, they say they have more legal and compliance work to undertake than ever before; and that the work is riskier too. Many General Counsel tell me that they are being required to reduce their overall legal budgets by between 30 per cent and 50 per cent. On the face of it, this is unsustainable. These clients from major companies and financial institutions are facing the prospect of an increasing workload and yet diminishing legal resources. Something surely has to give here. I call this problem the 'more-for-less' challenge—how can clients, working with their external legal providers, deliver more legal services at less cost? It is clear that clients' platforms are burning, even if partners proclaim the same is not so of leading law firms.

The more-for-less challenge is not just a conundrum for in-house lawyers. Small businesses face a similar dilemma. These traders do not have their own specialist in-house lawyers, and whenever they are in need of serious legal help, they must currently turn to external law firms. In these demanding times, however, many business people confess that they cannot afford lawyers and often have to run the risk of working without legal guidance. As for the consumer, although the law is central to all of our lives, dramatic decreases in public legal aid mean, effectively, that only the very rich or the very poor any longer have the means to afford the services of lawyers. Citizens face the more-for-less challenge too.

I believe the more-for-less challenge will underpin the next decade of legal service. This challenge will, I expect, irreversibly change the way that lawyers work. To be clear—no-one can sensibly contend that the foreseeable future will be one in which there will be less need for legal help. On the contrary, in an increasingly regulated and complex world, the demand for legal help is increasing. The big question is—how will this growing demand best be met? By traditional lawyers and law firms or by alternative providers?

Liberalization

The second main driver of change is liberalization. A little background should help here. In most countries, historically and generally speaking, only qualified lawyers have been permitted to provide legal services to clients, and, even then, only from specific types of organization (typically from partnerships). Laws and regulations have stipulated who can be a lawyer or legal professional, who can run and own a legal business, and what services they can provide. Different countries draw lines in different places. In England and Wales, the key concept is 'reserved legal activities', that is, work that only qualified legal professionals are permitted to undertake. In the US, the central concept is the 'unauthorized practice of law', which, by exclusion, carves out many more services as the sole province of lawyers. Whatever the disparity in local rules, the principles underlying the exclusivity of lawyers are similar in most jurisdictions; and the main justification is that it is in clients' interests that those who advise them on the law are suitably trained and experienced. Just as we would not want any Joe performing brain surgery on us, then, similarly, we should not wish that same Joe representing us in the courtroom.

But one big problem here is that this closed community of legal specialists does not seem to offer sufficient choice to the consumer. For decades, this has led critics and reformers to claim that the legal profession is an unjustifiable monopoly and that its practices are restrictive and anti-competitive. In turn, many have campaigned for a relaxation of the laws and regulations that govern who can offer legal services and from what types of business. This is a call for liberalization. (Note that liberalization is not the same creature as deregulation. Most campaigners for liberalization still want lawyers to be regulated; and indeed they want new categories of legal service providers to be regulated as well.)

In England and Wales, the call of these campaigners was answered as long ago as 2004, with the publication of an independent review, now known as the Clementi Report. Sir David Clementi (an accountant and not a lawyer) had been appointed by the Lord Chancellor to review the regulatory framework for legal services. Meeting and responding to concerns about restrictive practices in the legal marketplace, he recommended considerable liberalization. This led directly to the Legal Services Act 2007 which, amongst many other provisions, permits the setting up of new types of legal businesses called 'alternative business structures' (ABSs), so that non-lawyers can own and run legal businesses; it allows external investment, such as private equity or venture capital, to be injected into legal businesses by outside investors; and it lets non-lawyers become owners of law firms. The 2007 Act also permits law firms to move from being privately owned to issuing shares on a public stock exchange (globally, the first law firm to go public was Slater & Gordon in Australia in 2007).

This story of liberalization continues to unfold in England and Wales. The new ownership rules came into force in October 2011 and the licensing of ABSs began in March 2012. More than 1500 licences have since been granted. The Big 4 accountancy giants (KPMG, PwC, Deloitte, and EY) are amongst the recipients, one of many signals of their increasing interest in the legal market (see Chapter 17). Several long-established law firms (e.g., Reed Smith, Irwin Mitchell, and Weightmans) have been licensed as ABSs, as have innumerable smaller firms, alternative legal providers (see Chapter 17) and start-ups (Chapter 10). A few law firms have floated (e.g., DWF, Knights, and Keystone Law).

In combination, these developments are of profound significance and represent a major departure from conventional legal services. Not all of the moves were triggered directly by the Legal Services Act but this legislation—and here is the key point rather than the details of particular initiatives—is engendering a stronger entrepreneurial spirit in the legal market in the UK. Even where there is no formal liberalization, we are seeing a *liberation* from the constraints of narrow thinking about the way in which legal services can be delivered. There are new providers—new competitors—in the legal marketplace. No-one knows where this will eventually lead us. That is the nature of the market. All we can be sure of, I believe, is that major change is enveloping us. Private equity houses, venture capital firms, entrepreneurs, and other professional providers together are recognizing that the UK's £37 billion legal market is far from efficient and there are great opportunities for offering legal services in new, less costly, and more client-friendly and business-like ways.

Here is the nub. There are new players in the field, and they are not committed to traditional ways of working. They do not believe, for example, that legal work should be undertaken by expensive lawyers working in expensive buildings in expensive city centres. They do not insist, as many traditional lawyers still maintain, that legal work is best undertaken on an hourly-billing basis. They are not constrained by old ways of operating. They are passionate about change, and they are often better business managers than most lawyers, who tend to have had little training in the actual running of commercial concerns. How different the legal world will surely be when influenced over time by, say, the best retailers and tech companies, by the management methods of corporate boards, and with the backing of sophisticated external investors.

Market forces are sweeping through the legal profession in the UK and these will bring intense new competitive pressures for traditional law firms and greater choice for organizations and individuals who need legal guidance. The extent of the impact of liberalization on law firms is a matter of considerable current debate. Some major firms, for example, maintain that all of this is of relevance and threat only to High (Main) Street law firms that undertake high-volume, low-margin work (i.e., large quantities of legal work of modest value). These firms say, for instance, that they have no need for external investment. However, they should bear in mind that liberalization has helped entice all the Big 4 accounting firms to return to the English legal market with all the competition and investment capability that this will bring. The large firms may not feel they need extra cash to continue practising as they have in the past but it is not clear that they can comfortably afford to back new service opportunities, such as developing technology platforms (Chapter 6) or substantial artificial intelligence (AI) systems (Chapter 22), or setting up shared services centres for major clients (see Chapter 3).

Lawyers in countries yet to have been liberalized (which is most countries) will often dismiss the phenomenon of liberalization for a different reason—they regard this as a quirk of a small number of misguided jurisdictions. I anticipate, however, that when this liberalization gives rise to legal businesses and legal services that better meet clients' growing more-for-less challenge, then this will have a ripple effect around the world. General Counsel of global businesses who benefit from new forms of service in liberalized regimes will not unreasonably ask for similar service in their own countries. Law firms in traditional markets may find themselves at a competitive disadvantage, unable to raise funds for ambitious new ventures, for example.

Of course, whether and how other jurisdictions formally respond to the possibility of liberalization remains to be seen. Australia was first past the post. England and Wales have reformed emphatically. They are not alone. For example, the legal markets of New Zealand, Ireland, and Scotland have liberalized to some extent. In the US, many related questions have been under deep scrutiny by local bar associations, many of which seem to me quite bullish for change, and by the American Bar Association, conservatively in its Ethics 20/20 Commission (reporting in 2012) and more open-mindedly in its Commission on the Future of Legal Services (2016). Since then, there have been significant developments: in August 2020, Utah was the first state to authorize ABSs (albeit as a two-year trial); in the following month, the Supreme Court of Arizona announced that ABSs would be allowed on a permanent basis; while earlier that year, a task force of the State Bar of California made some encouraging noises in the same direction.

In the US, in my view, the dam is beginning to burst. I predict that, through the 2020s, after intense agonizing and various changes of direction, most major jurisdictions in the West (including much of the US) and many emerging jurisdictions too will have liberalized in the manner of England. And, even if they do not, liberalization in some countries will bring liberation in most others.

I remain broadly critical of jurisdictions that resist liberalization. I am afraid that those who resist this change in the name of protecting their clients are too often being disingenuous. Lawyers should survive and thrive because they bring value to society that others cannot, and not because they regulate others out of the field.

Technology

Much of my work over the past 40 years has been devoted to thinking and writing about the impact of technology (by which I mean information or digital technology) on lawyers and the courts. I have also advised innumerable law firms, in-house departments, and governments on this same subject. The legal profession has not generally been swift to embrace new systems but it is increasingly finding it impossible to avoid the technology tidal wave.

Technology is now pervasive in our world. Consider the number of users of mobile phones or devices (more than 5 billion), the Internet (also over 5 billion), of email (around 4 billion), and of Facebook (approaching 3 billion).

Think about data. In 2010, Google's then chairman, Eric Schmidt, said that every two days, 'we create as much information as we did from the dawn of civilisation up until 2003'. Every two days, on this view—*correction*: every hour or so today, because Schmidt was saying this more than a decade ago—we create more than 5 quintillion (5×10^{18}) bytes of data.

Digital technology is not a passing fad. On the contrary, courtesy of cloud computing, data storage and processing power are increasingly being made available as a utility, in the manner of water and electricity. And yet, many lawyers, in an untutored way, still tell me that technology is over-hyped. A few even still point to the bursting of the dotcom bubble and claim—based on who knows what—that the impact of technology is slowing down and that all recent talk of AI in the law will prove to be hot air. This is grotesquely to misunderstand the trends. Too few lawyers have heard of Moore's Law: not a law of the land but a prediction made in 1965 by Gordon Moore, the man who co-founded Intel. He projected then (to simplify) that every two years or so the processing power of computers would double, and yet its cost would halve. Sceptics at the time claimed that this trend would last for a few years and no more. In the event, it is still going strong and—given advances in material science, quantum computing, and chip architecture—it is likely to continue unabated for the foreseeable future.

In his formidable book, *The Singularity is Near*, the prescient Ray Kurzweil gives a practical illustration of the future consequences of Moore's Law, if it continues to hold. By 2050, according to Kurzweil, the average desktop machine will have more processing power than all of humanity combined. You can call me radical, but it seems to me that if we can see the day in which the average desktop machine has more processing power than all of humanity combined, then it might be time for lawyers to rethink at least some of their working practices. It is simply inconceivable that technology will radically alter all corners of our economy and society and yet somehow legal work will be exempt from any change.

Note too that this (literally) exponential growth in processing power is mirrored in most other aspects of technology, from the number of transistors on a chip, to hard disk capacity, computer memory, bandwidth, and more. Most arrestingly, in his contribution to the collection, *After Shock*, Kurzweil recently claimed that the performance of neural nets (the dominant technique underlying current AI) is now doubling not every two years but every 3.5 months, 'an increase of 300,000 fold in six years'.

The nature and role of technology is also changing. If you were a user of the Web in 1997 (when there were around 40 or 50 million users) you would have been the passive recipient of whatever information website providers chose to publish or broadcast in your direction. A decade later we moved into a new era—ordinary human beings (not computer specialists) were able to contribute and participate directly on the Web. Users became providers. Readers became authors. Recipients became participants. Users were able to contribute. We found radically new ways to produce information and to collaborate with one another, whether as bloggers, users of social networks, or contributors to shared, online resources such as Wikipedia and YouTube. The next chapters of the story, no doubt, will be devoted to Web3 (a decentralized version of the Web, based on blockchain—do not worry about the details) and to virtual reality and the metaverse, where users become much more fully immersed in online worlds.

It is exciting and yet disconcerting to contemplate that there is no finishing line for technology. Our machines and systems are becoming *increasingly capable*. Aside from the ongoing and radical changes in the underlying and enabling technologies, innumerable new applications emerge on a daily basis. It is bizarre to think that in a few years' time, our online lives will likely be dominated by systems that very few of us have heard of today, or indeed that may not yet have been invented. Consider

TikTok, the video hosting service. It was launched in 2016. Today, more than 1 billion people are users. And yet, even with that number, I always get the sense that lawyers are waiting for TikTok to take off. In resisting or denying the relevance of emerging systems, we are often witnessing a phenomenon that I call 'irrational rejectionism'—the dogmatic and visceral dismissal of a technology with which the sceptic has no direct personal experience. One of many examples here is 'online courts' (see Chapter 14). An embarrassing number of lawyers reject the very idea before taking the time to absorb what is actually involved, or seeing them in action.

I say again—our machines are becoming *increasingly capable*. This means that many of the tasks and activities that we have for long felt could only be undertaken by humans will, over time, be taken on autonomously by capable systems, or by less expert humans with the support of these systems.

A few examples help hit the point home. A long-standing favourite of mine is the achievements of Watson, IBM's computer system that competed—in a live broadcast back in 2011—on the US television general knowledge quiz show *Jeopardy!* Watson beat the show's two finest ever human contestants. This is a phenomenal technological feat, combining various techniques of AI, advanced natural language understanding, information retrieval, knowledge processing, speech synthesis, and more. While the remarkable Google retrieves information for us that might be relevant, Watson shows how AI-based systems, in years to come, might actually speak with us and solve our problems.

As remarkable as Watson, if not more so, is the progress that is being made in the field of machine learning. My favourite example is AlphaGo, a system designed by Google DeepMind to play the board game, Go. There are more possible moves in Go than there are atoms in the universe and AI specialists have for long doubted that any system, even in the medium term, could beat a great player. And yet, in early 2016, AlphaGo trounced the world's top Go player by four games to one. Using 'deep neural networks', the system was trained by a mixture of 'supervised learning' (based on past games of human experts) and 'reinforcement learning' (based on playing itself millions of times and improving from these solitary contests). One celebrated move by AlphaGo (the 37th move in the second game) was apparently described by a human champion as 'beautiful'. Disconcertingly for some, here is a system that displays characteristics that in a human being we would describe as 'creative' or 'innovative'. Certainly, many of the games AlphaGo played and the moves it made were well beyond the contemplation of those who designed the system. The sequel to this story is even more jaw-dropping. In 2017, the next version of the system, known as AlphaGo Zero, beat the original system by 100 games to 0. This new version was not trained on past data from human games. Instead, it was given the rules of the game and after three days of playing itself, it vanquished its predecessor.

Now contemplate GPT-3 (Generative Pre-trained Transformer 3), a language model that also uses deep learning technology. Developed by OpenAI and released in 2020, this natural language processing (NLP) system generates human-like text. It was trained on 200 billion words of text, and can read and write in a wide range of formats, on any subject. GPT-3 can perform tasks that we used to think were the exclusive preserve of humans. It can compose and summarize documents, answer questions, and take part in flowing conversation. It writes stories, creates poetry, devises screenplays. It can even draft an academic paper about itself. And it writes at a dauntingly high standard: in the original research paper, for instance, human beings were reported to be able to identify articles written by GPT-3 (as opposed to humans) just 52 per cent of the time—no better than random chance.

Finally, in my canter through examples of the increasing capability of machines, I must mention affective computing which focuses on the development of systems that can both detect and express human emotions. There is a system, for instance, that scan human smiles and, more accurately than humans, determine whether they are fake or genuine. Remember of these systems, as of all others, that they are only getting better. Remember too that most lawyers have never heard of Watson, AlphaGo, GPT-3, and affective computing.

Lawtech

What does it mean for the world of law that we are living at a time of greater and more rapid technological progress than humanity has ever witnessed, in an era in which our economic and social lives are being transformed by technology? As our machines become increasingly capable, it is wrong-headed to believe that legal practice and the administration of justice will be largely unscathed by technological advance. It is more likely, in what is one of the most information-intensive and document-intensive industries of all, that we will see fundamental, pervasive, and irreversible change.

In fairness, most lawyers do now accept that technology will affect them. But they expect its main impact will be in *automation* of what they already do today. They regard technology as a tool for streamlining, optimizing, and sometimes perhaps turbocharging their current ways of working. It is true that the first 65 years of lawtech (a term I use interchangeably with 'legal technology' or 'legaltech') were almost entirely devoted to automation, that is, to grafting technology onto traditional lawyering. Technology has been used over the decades to support and enhance conventional legal activity rather than to change it. What too many lawyers miss is the possibility and likelihood of *innovation*, a term I use in a specific way to refer to the way that technology can enable us to perform tasks that previously were not possible (or even imaginable). A mindset shift is needed here. When thinking about technology, the challenge for lawyers is not just to automate inefficient current working practices, which often gives us little more than 'mess for less'. The challenge, rather, for the 2020s and beyond, will be to innovate—to practise law, to offer access to justice, and to run our courts in ways that we could not have done in the past.

We must recognize, though, that many of these innovative technologies will be disruptive (see Chapter 6). This means they will not sit happily alongside and perpetuate conventional ways of operating. Instead they will fundamentally defy and change conventional habits. These ubiquitous, exponentially growing, innovative technologies will upend the legal world. Catalysed no doubt by the pandemic, they will change the client experience, shifting largely from in-person meetings in glass citadels or tired townhouses into rich, immersive online interactions. They will fuel a move from consultative legal advisory service to online products and solutions, and to embedding compliance and execution in the operations of businesses. They will underpin the development and delivery of online courts. They will deliver self-help systems for people who cannot afford lawyers (which is most people). These are elements of the story that is told in the pages that follow.

It is sobering to note, by way of orientation, that we are still at the foothills of technological change in law. The global spend on lawtech lies somewhere between $10 billion and $20 billion. Which may sound a lot, but the global legal market itself is creeping towards $1 trillion. Relatively speaking, much more is spent on technology in financial services and in healthcare. As important, if not widely discussed, is that (on my estimation) more than 90 per cent of that global spend on lawtech is by law firms; only a small fraction of the expenditure is by in-house lawyers for their own systems, and a tinier fraction still is on systems for direct use by users who are not lawyers (such as self-help services for citizens).

To drill down a little deeper, it is my experience from advising law firms that more 90 per cent of that lawtech spend is itself on back office systems (laptops, servers, data centres, software licences, and the rest). Unglamorously and prosaically, most lawtech today (90 per cent of 90 per cent, if you will forgive me) sits in the engine room of law firms. Most of the chatter about lawtech, in contrast, is about using technology to enhance and change client service. It is time to move from talk to action.

Many of the changes brought by technology, not least by social networking, should be familiar to younger members of the legal profession, as full-fledged members of the digital generation (which I define as those people who cannot remember a pre-Internet world). Interestingly, though, many young lawyers have not yet made the connection between their social use of technology and its introduction and potential in their working lives. Still fewer are engaged by AI and machine learning.

In conclusion, to pull together the strands of this opening chapter, I am suggesting that the more-for-less challenge, liberalization, and technology will together drive immense and irrevocable change in the way that lawyers work. There is a perfect storm brewing here. Liberalization and technology on their own would bring (and enable) reform but it is the more-for-less challenge, this imperative driven by demanding market conditions, that is and will continue to be the dominant force affecting the business of law in the 2020s.

2 | Impact of the Pandemic

In the Introduction and in the prefaces to the second and third editions of this book, I mention a prediction that I made when this book was first published—that the legal world would change more in two decades than it had in the past two centuries. Roughly speaking, I was referring to the period from 2012 to 2032.

Lawyers and prophets alike might well be uneasy if I now tried to argue that, because of COVID, we have already seen this extent of change, at least in the world of dispute resolution and in the working habits of lawyers. It is true, nonetheless. I accept there is something unsavoury about claiming any success because of COVID. And I concede that I ought not generally to take credit for events that I had not anticipated. On the other hand, many of the changes wrought by COVID are ones that I did indeed foresee. COVID, in effect, accelerated numerous technological advances that I have discussed in my work over the years.

I could leave it there but to do so would be to oversimplify—it is too soon to draw definitive conclusions about remote working in law; we have not gathered and analysed sufficient data about the impact so far, and there is evidence to suggest that COVID has decelerated as well as accelerated the uptake of legal technology.

Initial Response

Even some techno-sceptics in the law are now saying they will never return to their old ways, that the virus catalysed their adoption of technology. Lawyers are routinely meeting clients by video, law students are learning online, and many court hearings are being conducted remotely, all to an extent that would have seemed ludicrous if proposed at the start of 2020. Does this mean the legal profession has irreversibly and comprehensively embraced technology? The advance has been remarkable but there is a danger here of exaggeration.

We should not mistake the temporary systems that were stitched together in haste—putting lawyers, students, and judges on, say, Zoom—with the full-scale industrialization of these early efforts. To install secure and robust systems at scale will require formidable further investment. Whether in the law office, law school, or law courts, the move to sustainable, long-term remote working will call, for example, for new management structures, greater attention to health and safety compliance, superior connectivity, and better equipment for the millions who look likely this decade to be stationed frequently at home. For some, of course, their home environment may not be suitable as a permanent or semi-permanent base. The move will not work for everyone.

Meanwhile, in law as elsewhere, old habits die hard. It may not be modish to confess to this in public, but some if not many reactionary lawyers, law professors, and judges have been unobtrusively hunkering and hankering—hunkering down until the viral storm passes, while hankering after their old ways of working.

Unscheduled Pilot

In a sense, we have been living through a major unscheduled pilot, an experiment in the use of a range of technologies that have been keeping legal and court services operational. In many cases, this has been a proof of concept. COVID created a sense of urgency. It opened and changed many minds. It gave rise to pragmatic, workable solutions and in so doing confirmed that best practice can be the enemy of good practice. In the past, systems for lawyers and the courts have been over-specified and over-engineered and have taken much too long to implement.

At the same time, we have regarded in-person engagement as the defining gold standard. In reality, however, we cannot afford the alleged best on all occasions—the face-to-face service. The pandemic has helped us to recognize that good enough is frequently good enough and certainly better than nothing.

But these are very early and tentative conclusions. If we indeed regard the response to the pandemic as an experiment, we should be as rigorous and systematic in analysing its results as our scientists were in their searches for treatments and vaccines. We must gather data about volumes, timings, technologies, applications, users and their experiences. And we need new methods for evaluating the data and for future piloting. We need in law the equivalent of the randomized controlled trials that are regarded as the best techniques for assessing and comparing medical interventions.

In truth, in the flurry of keeping organizations operational, the collection of data about the impact of the technology has been patchy. Neither law firms nor court services know enough about what has worked well and what has not in our remote offices and courts. When we do manage to capture enough data about successes and shortcomings, we can then use this—rather than anecdote and speculation—as the basis of informed decisions about what should be preserved post-COVID and when we should revert to traditional methods. In turn, our policy-making about radical change in courts and our strategy formulation in law firms can be genuinely evidence-based.

The data gathered should be the basis of innovative thinking in law and justice. And goodness knows, we need some new, creative ideas. Long before the virus appeared, our justice system was creaking—unaffordable, antiquated, and invariably unintelligible to those who are not lawyers. The pandemic brought our challenges into sharper relief.

With all of that said about data, we should also remember that the status quo—how our legal profession and our courts currently operate—is not an evidence-based option that we have consciously chosen. It is simply what we have become.

Disruption on Hold

Let us also be clear, however, that working from the kitchen table does not constitute the long-hailed disruption of the business models of lawyers. Uncontestably, we have witnessed an accelerated deployment of some technologies, but most of these have essentially propped up our conventional ways of applying and delivering legal knowledge—through in-person engagement.

This is not, in my view, the seismic transformation that awaits us, which is when many of the activities of lawyers and the outcomes they deliver are themselves undertaken by increasingly capable systems operating autonomously or in support of non-expert users. That transformation, in many ways the focal point of this book, has surely been decelerated by COVID. For most legal professionals, for example, artificial intelligence was put on hold during the pandemic, while we urgently put in place methods of communicating and collaborating when unable to meet physically. The more ambitious and transformational technologies were put on back burner by most legal organizations. In summary, in the language of Chapter 1, the pandemic accelerated automation and decelerated innovation in the world of law.

Post-pandemic, to overcome the many shortcomings of our current legal and court service (as discussed across Parts One and Two), the challenge will be to move beyond grafting technologies onto traditional operating practices. We will turn again to the development of systems that will replace rather than automate our old ways of working.

In all, in the grand arc of transformation in the legal market, I regard the pandemic not as a driver of change in its own right. To be sure, it has affected the pace of one of my three fundamental drivers—technology—as an accelerator (and decelerator), but not the general course of technological advance. The pandemic's effect has been shocking and its impact far-reaching, but I expect history to confirm that it did not change the overall direction of travel set out in Chapter 1.

3 | Strategies for Success

The three drivers of change introduced in Chapter 1 are urging law firm leaders around the world to contemplate opportunities and threats that the legal market has had little reason to confront in the past. With clients under cost pressures and the business environment changing rapidly, prudent law firms everywhere are trying to develop compelling responses to the new market conditions. In other words, law firms are spending much time and effort in thinking through their strategies for the next few years and beyond.

Charging Less

It might be thought that the best way to meet the more-for-less challenge would be for law firms simply to charge less. For businesses that enjoyed almost uninterrupted yearly growth in profit and turnover for the 20 years leading up to 2007, and then again for more than a decade after the downturn, the suggestion of charging less is not normally greeted with unbridled enthusiasm. Nonetheless, law firms like to show willing and so many propose 'alternative fee arrangements' (sometimes known as AFAs) to their clients. The 'alternative' that lawyers have in mind is to 'hourly billing' which has been the dominant way of charging for legal services since the mid-1970s. In truth, hourly billing is not simply a way of pricing and billing legal work; it is a mindset and a way of life. Lawyers charge for their time—for their input and not their output. And, until not long ago, most clients have seemed comfortable with this approach.

The shortcomings of hourly billing are well illustrated by an old anecdote involving my daughter. When she was 12, she asked me for a summer job. I needed some administrative work carried out and she agreed to take on the task. She asked me how much I intended to pay her and I responded, unreflectively, that I thought I would pay her a certain amount per hour. She thought about that for a few seconds, smiled, and then said: 'Well, I'll take my time then.' If a 12-year-old can see the shortcomings of hourly billing, then it puzzles me that major international corporations cannot also see the problem here. Hourly billing is an institutionalized disincentive to efficiency. It rewards lawyers who take longer to complete tasks than their more organized colleagues, and it penalizes legal advisers who operate swiftly and efficiently. All too often, the number of hours spent by a law firm bears little relation to the value that is brought. A junior lawyer who expends 50 hours on a task can sometimes provide much less value than half-an-hour of the work of a seasoned practitioner (drawing on his or her lifetime of experience).

The dominant culture in so many major commercial firms, however, is still for lawyers to churn out as many chargeable hours as possible. Underlying this practice is a business model for professional firms that has ruled for several decades—the ideal, in theory and practice, is to have a pyramidic structure at the top of which is the equity partner (the owner) of a law firm, beneath whom are junior lawyers whose efforts bring far more revenue to the firm than they are paid as salary. On this model, the broader the base of the pyramid, the more profitable is the firm. And so, in major US firms, for example, many associates are expected to work well over 2,500 chargeable hours each year, a set-up that ensures great profitability for law firms but one with which clients are increasingly disillusioned.

In passing, I might add a word or two about rates and incomes. In those large commercial firms where partners' hourly rates comfortably exceed, say, £1,000 per hour and their associates' charges are about half of this, this yields very significant profits for these partners. There are north of 100 firms in the world in which many of the partners earn over £1 million per year and in some, their take is numerous multiples greater than this. Many of these partners confess that when they entered the law they never dreamt of such incomes and that they had not chosen the law as a career because it would be well remunerated. In contrast, many high-powered law graduates today enter the law precisely because of the promise of considerable wealth. They may be disappointed. Although a handful of these global practices are likely to continue earning very substantial incomes, it may well be that the golden era for many law firms has almost passed. Over time, the more-for-less challenge will drive down profitability.

The scales of income just mentioned understandably give rise to media and public characterization of lawyers as 'fat cats'. However, the overwhelming majority of lawyers around the world earn much more modestly. In most large jurisdictions, approximately 30 to 40 per cent of the law firms are run by sole practitioners and about 75 per cent have four partners or fewer. In these practices, the profits are considerably lower, in line with senior public sector workers rather than private bankers.

Alternative Fee Arrangements

Returning to the vexed issue of charging, many law firms, as said, have sought in recent years to meet clients' demands for lower fees by proposing methods of charging that are not time-based. There has been an upsurge of proposals for work being undertaken on a fixed cost basis or on a capped basis (where an upper limit to the fees is agreed). Others have gone further and put forward more exotic approaches such as 'value billing', which involves, in a variety of ways, charging for the value of the work undertaken rather than the time expended; or, a variant on this, charging for time and cost saved rather than time spent.

These proposals have generally been prompted by in-house lawyers who, under cost pressures, have formally invited law firms to submit 'new' or 'innovative' suggestions for the pricing of their services. These requests have often been made as part of a broader process of selection of what are known as 'panels' of law firms. A panel, essentially, is a group of preferred firms. The selection process is quite formal, undertaken through byzantine documents, entitled RFPs (requests for proposals) or ITTs (invitations to tender). Increasingly, in-house lawyers have also been working alongside, or been displaced by, professional procurement people who are more experienced in driving down the cost of external suppliers.

There is much to be said both for and against these panels and procurement professionals, but, for now, the crucial point to grasp is that this competitive tendering process does not seem to be yielding the savings that clients require. Alternative fee arrangements seem to be failing to deliver significant savings for clients for at least two reasons. The first is that most AFAs are derived from hourly billing thinking—in calculating fixed fees, for example, the starting point of many law firms is the amount that would have been charged on a conventional, hourly billing basis. Fixed fees, therefore, often represent but a slight variation on hourly billing. Second, and more importantly, very few firms when proposing AFAs do so with the intention of becoming less profitable; and so, if they do not propose to change the way they work (and rarely do they), then the alternative fee proposal is often little more than a repackaging of the original (too costly) proposition. The feedback I am hearing and the private research I have examined continue to suggest that competitive tendering and the resulting proposals for alternative fee arrangements are delivering to clients an overall reduction in the cost of legal services of about 10 per cent. Whether they are major organizations or consumers, the harsh reality for clients who need to slash their legal budgets in half is that pricing differently will not be sufficient to meet their more-for-less challenge. I believe it is now necessary to move from pricing differently to working differently.

The Two Winning Strategies

In my view, there are only two viable strategies available to the legal world to help it cope with the more-for-less challenge. I call these the efficiency strategy and the collaboration strategy. In short, the efficiency strategy maintains that we must find ways of cutting the costs of legal service, while the collaboration strategy suggests that clients should come together and share the costs of certain forms of legal service. The efficiency strategy is likely to be favoured over the next few years, whereas the collaboration strategy will come to dominate in the longer term.

Many law firm leaders, when they hear me speak of the efficiency strategy, agree immediately that legal costs need to be reduced. They may then go on to discuss how their overheads should be trimmed, often by spending less on back-office functions such as technology, marketing, and human resources. It may be that such measures are appropriate in running a leaner machine, but these are not the cost reductions to which I am referring when I advocate the efficiency strategy. Instead, my claim is that the cost of lawyering itself has become too high. Most clients tell me that they do not mind paying significant rates for experienced lawyers but they do object, with increasing indignation, to paying, for example, high hourly rates for relatively junior lawyers to undertake what they perceive as routine and repetitive work. This is the crux of the matter.

In every legal business I visit or advise, I find significant amounts of work being undertaken by young lawyers that is administrative or process-based. The work requires more process than judgment, procedure instead of strategy or creativity. Examples are document review in litigation, due diligence work, basic contract drafting, and rudimentary legal research. Here is the great opportunity for change. It is to identify work that can be made routine and undertaken more efficiently, whether by less qualified, lower-cost human beings, or through computerization. This leads us naturally down a path towards the 'commoditization' of legal work (see Chapter 4) and to what I have termed the 'decomposing' and 'multi-sourcing' of legal work (Chapter 5). These are not fanciful theoretical notions. They are a principal preoccupation of most of the in-house lawyers with whom I now meet and of many law firm leaders too.

As for the collaboration strategy, this is more radical and, at first sight for many lawyers, may seem implausible. The idea, once more, is that to meet the more-for-less challenge, clients can and will come together and share the costs of certain types of legal service. This strategy can be pursued in conjunction with, or instead of, implementing the efficiency strategy. The most dramatic example of the collaboration strategy is one I have advocated for some time for major banks. It applies to their work in regulatory compliance. Major banks spend many hundreds of millions of pounds each year on compliance. Many of these financial institutions operate in well over 100 countries, each with different legislation and regulations, and each requiring not only compliance with their respective rules but also regular submissions of documentation and forms to their regulatory bodies. Keeping up to date with new regulation and changes in old regulation, educating tens of thousands of people on their obligations, understanding the local practices and preferences of regulators, introducing standard processes for supporting the preparation and submission of documentation—these are the tasks facing compliance specialists.

My simple contention for some years has been that some banks could come together and share the costs of undertaking many of the compliance jobs that they have in common. This would not be appropriate, of course, for compliance tasks that are sensitive, confidential, or competitive; but much compliance work is administrative and non-competitive and the duplication of effort across the banking world is massive and unnecessarily costly. My suggestion, therefore, is that banks club together and set up, for example, shared services centres, which would help them to undertake at least some compliance activities at vastly reduced cost. This would be going a step further than syndicating legal advice, as is sometimes already done. For law firms that currently benefit from advising each bank in turn on their compliance work, this 'compliance process outsourcing' (as I call it) would be a grave development. No longer would they be able to recycle much of their work across their various clients. Instead, clients would collaborate with a smaller number of supporting firms. I expect that one or two law firms would enjoy great commercial success if directly involved in supporting the collaborating groups of banks.

Clients can also collaborate in the development of systems. An early case study here was Rulefinder, an online legal risk management tool developed by the international law firm Allen & Overy. This service offers help with the rules and practices relating to international shareholding disclosure. This is a complex and frequently changing area of regulation that affects all major financial institutions. Innovatively, when the system was first developed, six leading banks came together and collaborated with Allen & Overy and so shared the costs of producing the system.

But the collaboration strategy is not just for large financial institutions. Some years ago, in England, for example, the in-house legal departments of a number of local authorities convened and in a similar way shared the costs of common legal work. This philosophy could equally extend to small businesses and individuals—new-look legal businesses will no doubt spring up to serve communities of legal users rather than individuals or organizations on their own.

I first floated the idea of the collaboration strategy in 2008, in my book, *The End of Lawyers?* The suggestion was met with tentative approval by in-house lawyers and sceptical incredulity by leading law firms. In my recent travels (often by video) in North America and continental Europe, I have heard General Counsel speaking with growing interest about this possibility, supported by purpose-built online platforms. Quietly but steadily, the collaboration movement is gathering pace.

4 | Commoditizing the Law

Central to the efficiency and collaboration strategies, as introduced in the previous chapter, and also to the general idea of working differently, is a term that is vile and yet vital—that of 'commoditization'. This has become a rather overused notion in the legal world and, unhelpfully, it is a word that is often bandied about with little precision. When many lawyers speak of commoditization, they are prone to do so in bleak and dismissive terms—commoditized legal work, it is intoned with deep regret, is work from which we can no longer make money. The thrust here is that work that was once handcrafted can now be made routine and disposed of quickly with little need for lawyers' intervention. In contrast, from the client's point of view, this shift towards routinization tends to be a good thing, because it leads to much lower expenditure on legal work.

A False Dichotomy

Commoditized legal work (loosely so called) is often distinguished from what I term 'bespoke' legal work. I have used the word 'bespoke' for many years but have come to realize that, beyond England, it sometimes requires some explanation. Think about clothing for a moment. A bespoke suit is an outfit that has been customized, made to measure, and tailored for the precise contours and topography of its owner. It is handmade, handcrafted, and cut specifically for one individual alone. By analogy, I believe that many lawyers regard legal work as highly bespoke. Their client's circumstances are unique and each requires the handcrafting or fashioning of a solution, honed specifically for the individual matter at issue. This is the conception of legal problem-solving that is impressed upon law students in many law schools, where it seems that all problems put before them have features so distinctive that they could require the attention of the Supreme Court. It is also a model of legal services that is found in our literature and theatre, when lawyers look assiduously for smoking guns and loopholes.

I take the view that regarding legal work as bespoke in nature is an unproductive—if often romantic—fiction. I accept that some legal issues that arise do call for the application of acute legal minds and the handcrafting of tailored solutions. But I believe much less legal work requires bespoke treatment than many lawyers would have their clients believe. More than this, I contend that deploying bespoke techniques in many instances is to adopt cottage-industry methods when mass production and mass customization techniques are now available to support the delivery of a less costly and yet better service.

A further source of confusion here is the oversimplification in thinking which concludes that work is either bespoke or in some vague sense commoditized. This dichotomy urges many lawyers to insist that if they are to avoid non-profitable commoditized work, they must then focus only on bespoke endeavour. They believe these are the only two options.

The Evolution of Legal Service

I maintain this binary distinction between bespoke and commoditized legal work is a false dichotomy and that legal services are in fact evolving through four different stages which I call bespoke, standardized, systematized, and externalized, as depicted in Figure 4.1. (Readers of the first

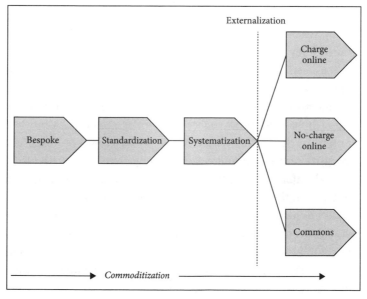

FIGURE 4.1 *The evolution of legal service.*

edition of this book will recognize that I have moved away from the five-stage model presented there. The revised model better reflects the developments we are actually seeing in lawtech and, I hope, more helpfully explains the concept of commoditization.)

In practice, much of the work of good practitioners is not undertaken in a bespoke manner. To be sure, and I want to stress this again, difficult problems do arise that undoubtedly require bespoke attention; but, far more frequently, lawyers are asked to tackle problems which bear a strong similarity to those they have faced in the past. Indeed, one of the reasons clients select one lawyer over another, or one firm over another, is precisely that they believe that the lawyer or firm has undertaken similar work previously. Most clients would be horrified to think, especially if they are being billed on an hourly basis, that each new piece of work they pass to law firms is set about with a fresh sheet of paper and embarked upon from scratch. On the contrary, clients expect a degree of standardization.

Take an employment contract as an example. If a bespoke approach were adopted, each employment agreement would be drafted afresh, starting with a blank canvas. However, unless the circumstances of the employment arrangement were particularly unusual, informed clients would expect standardization in two forms. First, they would imagine some form of standard process would be in play—perhaps a checklist, playbook, or procedure manual. Second, they would anticipate that their lawyers would use some kind of standard template or precedent as a starting point. In most reputable law firms, this kind of standardization, both of process and substance, is widely embraced. Clients have no interest in paying for re-invention of the wheel.

But the evolution of legal service does not stop at standardization. With the advent of technology, a further step can be taken—that of systematization. This can extend to the computerization of checklists or procedure manuals into what are known as workflow systems. These are commonly used in the insurance industry where there is automation of high-volume, often low-value, tasks and activities. Where there are many tasks, activities, and people involved, and yet the process can be proceduralized, automated workflow can greatly enhance the efficiency of legal work. Systematization can also extend, however, to the actual drafting of documents. To use the employment contract example again, document automation is a technique that requires users to answer a series of questions on screen (for example, the name of the employee, the starting date of employment, salary, and so forth) and after completion of an online form, a relatively polished first draft is compiled and emerges. The dominant underlying technology for this has been in existence since the early 1980s—it is a kind of rule-based decision tree, so that answers to particular questions cause a paragraph or sentence or word to be inserted or deleted, as the case may be. Document automation tends to have the added advantage that the user answering the questions need not be a legal expert or even a lawyer.

I know of one firm that systematized its document drafting internally more than a decade ago and claimed that this new efficiency was a key differentiator for them in the market. However, one savvy client perceptively inquired: if drafting employment contracts, at least for the majority of employees, involves no more than completing a form online, then why can this not be done directly by the human resources department within the client organization? This line of thinking leads naturally to the 'externalizing' of legal services. This occurs when lawyers pre-package and make their experience available to clients on an online basis. It offers an entirely new way of tapping into lawyers' expertise. Different ownership and charging models can be used when externalizing. The service can be made available on a chargeable service (the favoured option for commercial law firms), sometimes at no cost (an approach preferred by government bodies and charitable organizations), and occasionally but increasingly on a commons basis (in the spirit of Wikipedia and the open source movement).

I now think that this entire transition—away from bespoke service towards externalization—can usefully be termed the 'commoditization' of legal service. On occasions, some lawyers refer to what I would call 'standardization' as 'commoditization' whereas others equate commoditization with no-cost externalization. However, even when law firms or other providers charge for access to their online services, this can still mean dramatically lower costs of service for the client, while for the law firm it offers the opportunity to make money while they sleep—this is a radical departure from the hourly billing model, because the lawyers' expertise is used without any direct consumption of their time.

For many years, I have advised the tax practice of Deloitte, and here we find an early case study of commoditization. From around 2000, they progressed along an evolutionary path in respect of their tax compliance work (helping clients to prepare and submit their corporate tax returns) that is similar to the one introduced in this chapter. In the beginning, this was a handcrafted activity but they moved steadily along the spectrum and, in the UK, they distilled the collective expertise of around 250 of their tax specialists into a system for clients to use directly. In externalizing their tax knowledge in this way and selling it, they fundamentally changed their business model. They created a service that they intended would be of lower cost to clients and, because they have so many users, more profitable for Deloitte than the traditional bespoke offering. Interestingly, Deloitte sold that service to Thomson Reuters in 2009.

From the client's point of view, the arguments in favour of moving from left to right on my evolutionary path are compelling—as we move from left to right, the cost of legal service comes down, the price becomes more certain, the time taken to complete work reduces, and the quality—surprisingly for some—goes up (the collective expertise of many professionals invariably outclasses even the most talented bespoke solo performance).

Many lawyers respond dismissively to the idea of externalizing legal services as online offerings. They say that they did not go to law school to package or externalize their knowledge, they are not publishers, and they are certainly not software engineers. I look at the world very differently. I think if we can find new, cheaper, more convenient, and less forbidding ways of delivering legal services, then we should be adapting the way we work and adopting these new techniques. Our focus should be on helping our clients to meet their formidable more-for-less challenge rather than obstinately holding on to outdated, inefficient working practices.

There is no denying, however, that my model suggests, rather ominously for lawyers, that some legal resources will become readily available online at no cost, perhaps even as a shared resource to which anyone can contribute and from which anyone can draw. I acknowledge that lawyers will not benefit commercially from the commoditization of legal services in these ways, but I urge that these forms of externalization will be fundamental in radically increasing access to justice for those who cannot currently afford legal services.

5 | Working Differently

I am anxious to add one point of clarification to my discussion of the evolution of legal services in the previous chapter. What I am not saying is that for any piece of legal work—say, a deal or dispute—the question that arises is as follows: in which of my six boxes does that legal matter sit? I am saying something subtler than this, namely, that for any deal or dispute, no matter how small or large, it is possible to break it down, to 'decompose' the work, into a set of constituent tasks. And it is in respect of each of these tasks, not the job as a whole, that one can ask: what is the most efficient way of undertaking this work, and to which of the six boxes should the tasks be allocated?

If my first major point in this book is that the legal market faces the more-for-less challenge, then my second is that legal work can be decomposed and sourced in different ways.

Decomposing

Legal engagements such as deals and disputes, I am saying, are not monolithic, indivisible professional engagements that must all be sourced and undertaken in one way. Instead we can decompose (others would say 'disaggregate' or 'unbundle') work into various tasks and should discharge each, I propose, in as efficient a manner as possible. None of what I say is to betray quality. Rather, my claim is that there are ways of undertaking individual legal tasks that will deliver quality as high as conventional legal service (and sometimes even higher) but at far lower cost.

When I speak at conferences about decomposing legal work, I am often met afterwards by a lawyer who will, quite amicably, tell me that my remarks were enjoyable and will agree that the legal world is in need of a considerable shake-up. The lawyer will go further and concede that what I say about commoditization and decomposition applies to every area of legal practice . . . except one. And the lawyer will then tell me why it is that what I say about working differently does not apply to his or her own area of legal work. Particularly insistent are litigators who will maintain that every dispute is unique and that there is no scope for decomposing and the rest. This reflected my experience of the 1990s, when I worked for many years with a leading firm of litigators (Masons, now Pinsent Masons). It was true, at the time, that dispute work was not decomposed and it was common, for the large construction and technology disputes in which we specialized, for almost all aspects of the work to be handed over to the firm in their entirety. However, I have since come to see that not all of the tasks that we and other firms then undertook are any longer best discharged by law firms. And so to the sceptics by way of demonstration, I suggest that the conduct of litigation can be divided into the ten tasks that are laid out in Table 5.1. I am not suggesting that this is the only

TABLE 5.1 Litigation, decomposed

document review
legal research
project management
litigation support
(electronic) disclosure
strategy
tactics
prediction
negotiation
advocacy

way of decomposing litigation but I hope it gives a sense of my approach.

Over the years, formally and informally, the question I have asked litigators in the finest of the world's law firms is this: which of these ten tasks are you *uniquely* qualified to undertake? In the UK, the answer to this query has invariably been 'two' tasks (strategy and tactics) and, in the US, the answer has tended to be 'three' (strategy, tactics, and advocacy). And for these two or three tasks, clients will continue to want the direct advice and guidance of skilled lawyers. However, I am increasingly hearing from General Counsel that alternative providers can now take on the remaining tasks at lower cost and to a higher quality than traditional law firms.

Take, for example, document review. Traditionally, junior lawyers are deployed by law firms, at significant hourly rates, to work through large bodies of documents (sometimes many millions of them), often simply to index them or to impose some very basic legal classification. Leaving technological solutions to one side for now, manual document review can be outsourced to third party specialist providers, in low-cost countries such as India, and undertaken to a higher quality for around one-seventh of the cost.

Consider a further illustration, that of project management. Many litigators confide in me that they are no longer lawyers; they are now project managers. I sometimes inquire about the extent of their training in project management and am often told, with a straight face, that they went on a two-day training course three years previously. I tend to quip in response that if a project manager said to a lawyer that he or she was now a lawyer, having undertaken a three-day training course in the law, that project manager would be dismissed as misguided. Project management is a significant discipline in its own right, with its own techniques, methods, systems, and degree courses. When I look inside major accounting firms, consulting practices, and construction companies, I find sophisticated project management. In law firms, on the other hand, project management often seems to involve little more than buying some new lever-arch files and cracking open a new pack of yellow stickers. It is our collective arrogance as lawyers that we feel we can take on a neighbouring discipline over a weekend. We cannot. And clients now recognize that they will find the best project managers not in law firms but within other providers. I strongly believe that project management will be central to the successful conduct of large-scale disputes (and deals) in the future. But if lawyers are not sufficiently trained in this discipline, competitors from other professions and sectors will undertake this work in their stead.

TABLE 5.2 Transactions, decomposed

due diligence
legal research
transaction management
template selection
negotiation
bespoke drafting
document management
prediction
legal advice
risk assessment

I can undertake similar analysis of each of the other tasks in litigation for which law firms are no longer uniquely qualified. Many of these tasks are routine and repetitive, largely administrative, and can now be sourced in different ways. Equally, I can subdivide transaction work into an analogous list of tasks, as in Table 5.2 (again, not put forward as definitive but included simply to give a taste of what I have in mind).

Alternative Sourcing and Multi-Sourcing

When I say that pricing differently is not enough and that lawyers must move towards working differently, I have in mind the adoption of one or more alternative ways of sourcing legal work. In the past, when confronted with a legal job, a client had a simple choice: to undertake it internally or to pass it out to an external law firm (or perhaps a blend of the two). The legal world has changed, so that new alternative sources of legal service are now available. I have identified 16 ways of sourcing legal work, as laid out in Table 5.3. In this introductory book I can give but a hint of each.

In-sourcing is when lawyers undertake legal work themselves, using their own internal resources. This could be, for example, when an in-house legal department decides to conduct all of its negotiation and drafting internally, without any external advice or assistance.

TABLE 5.3 Sources of legal service

in-sourcing
de-lawyering
relocating
off-shoring
outsourcing
subcontracting
co-sourcing
near-shoring
leasing
home-sourcing
open-sourcing
crowd-sourcing
computerizing
solo-sourcing
KM-sourcing
no-sourcing

De-lawyering is my inelegant term for the process by which a legal task is handed over to a non-lawyer to discharge. Many tasks do not require the expertise and cost of qualified lawyers and can be taken on by other skilled and knowledgeable individuals now working in the legal sector.

Relocating involves an organization moving some of its legal work to less costly locations, but still within countries in which the main business already has a presence. An early illustration here was the US-based international law firm Orrick, which established a global operations centre based in Wheeling, West Virginia.

Off-shoring is the transfer of legal work to countries in which labour and property costs are lower. Many large banks have off-shored some of their legal activities in this way—for example, to India and Malaysia—to places where they have already moved other functions, such as their call centres or their finance functions. On this model, the off-shored legal resource remains part of the bank.

Outsourcing, in contrast, entails the conduct of legal work by a third party provider. This is often referred to as 'legal process outsourcing' or 'LPO'. Routine legal tasks, such as document review, are handed to these specialist support companies, which, again, are usually in low-cost locations.

Subcontracting is an option open to law firms. On this approach, legal work is passed to other (usually smaller) law firms, which carry much lower overheads. Thus, several large London-based law firms subcontract work to English qualified lawyers working in South Africa and New Zealand, while others engage lower-cost regional firms within the UK. Subcontracting can halve the costs of certain legal tasks.

Co-sourcing occurs when organizations collaborate in the delivery of some legal service, often through some shared services facility. Powerful illustrations of this, as noted in Chapter 3, are the cooperation of the in-house legal departments of local authorities in England and the plans of some banks to use common facilities for the conduct of their compliance work.

Near-shoring is similar to off-shoring but the work is carried out in a neighbouring, low-cost jurisdiction that is in a closer time zone to the law firm or in-house department that is parcelling out the legal tasks. Allen & Overy and Herbert Smith Freehills, leading international law firms, have both near-shored by setting up facilities in Belfast, in Northern Ireland, for the disposal of routine legal work.

Leasing is the engagement of lawyers for limited periods and often on a project basis. These lawyers do not belong to conventional law firms. Instead, they are made available through agencies that manage their placement. Axiom was a pioneering example of such an agency. Founded in 2000, the original premise of this business was to lease lawyers largely to corporate clients, often to help them to meet peaks in demand. This approach has proven particularly useful for in-house departments that downsize, because they will periodically need to boost their own capability and Axiom lawyers can be about half the price of those from conventional firms. Significantly, though, several law firms in England have set up similar leasing facilities—Bryan Cave Leighton Paisner (in 2008) with its Lawyers on Demand service, Pinsent Masons (in 2011) with Vario, and Allen & Overy (in 2013) with Peerpoint.

Home-sourcing embraces legal talent that is not currently in the mainstream legal workplace and yet is available, often on a part-time basis, from lawyers who choose to work from their own homes. During and after the pandemic, of course, home-sourcing has become commonplace. Enabled to a large extent by ever-improving communications and collaboration technologies, lawyers who work to a large extent from home (whether on an employed basis or as independents) are able to join and use the networks of law firms and in-house departments with whom they are working. Home-sourcing not only enabled the legal profession to survive during the pandemic but it offered a taste of an entirely different lifestyle, not least for parents who wish both to work and to be available for their young children for much of the day.

Open-sourcing is the provision, at no charge, of all sorts of legal materials (standard documents, guidelines, procedures, opinions, case studies, practical experience, and more) on publicly accessible websites. This is likely to be most effective if organized in the form of a wiki (an online resource that any users can edit and add to).

Crowd-sourcing involves harnessing the collective talents of large groups of individuals who make some of their time available to undertake certain categories of legal task. On one approach, for instance, a legal problem might be broadcast to a large, unknown group of volunteers. And these volunteers—the crowd—respond with their proposed legal solutions. In law firms, practitioners often pop their heads around doors and ask colleagues in relation to an issue they are handling, 'has anyone seen one of these before?' In the future, lawyers and clients will be able to ask similar questions of large bodies of Web users. It will be commonplace for people to ask questions online and for the answers given by lawyers or perhaps the recipients of past advice to be shared with other users.

Computerizing, although the term is beginning to sound dated, is a wide category of sourcing which I take to include the two categories of systematizing and externalizing, as introduced in Chapter 4. In general terms, computerization refers to the application of technology to support or replace some legal tasks, processes, activities, or services.

Solo-sourcing is the engagement of individual specialists, such as law professors or (as is common in England) barristers to conduct specific, decomposed packages of legal work. Illustrations of this are research conducted by academics and opinions written by KCs.

KM-sourcing is the use of a variety of techniques from the field of knowledge management to reuse content, process, know-how, sources, ideas, and much else that have been captured from daily practice, and preserved precisely for later recycling. Clients often expect and invariably prefer lawyers to use materials that have worked successfully in the past in similar circumstances.

No-sourcing is my final category, and is the option of choosing not to undertake a legal task at all, on the informed view that the task itself is not sufficiently high risk to merit any form of legal sourcing. As an illustration, in-house lawyers often take a view on certain portions of legal work and decide that the time and expense that they would require is not commercially justified. They often find this easier to do when work has been decomposed in the manner described earlier in this chapter.

Although each of these 16 techniques, if deployed in isolation, can provide powerful alternative ways of sourcing legal tasks, it is short-sighted to view them as distinct options. In the future, for any substantial piece of legal work, I expect that it will become common practice to decompose the matter in question into manageable tasks, to identify the most efficient way of sourcing each task, and to adopt several of the alternative approaches in combination. I call this 'multi-sourcing'. Thus, for a particular deal or dispute, a few if not many sources might contribute to the final product. To achieve this, we may find it useful to apply production-line or manufacturing mentality and methodology to the delivery of legal services: using, for example, just-in-time logistics and global supply chain techniques (underpinned by technology). On this model, one individual organization—a law firm or perhaps a new-look legal business—is likely to take over all responsibility for the delivery of the completed, multi-sourced service (as a main contractor will do in a building project).

I am not suggesting that multi-sourcing is straightforward. Experience suggests that law firms and legal departments find it hard in practice to disaggregate and farm out to alternative providers. These lawyers would benefit from improved method. Nor am I suggesting that all multi-sourcing must involve multiple suppliers—some clients are nervous about having too many cooks but like the idea of individual legal providers multi-sourcing inside their organizations. On this approach, law firms or alternative providers are encouraged to decompose work but rather than to collaborate with others, to multi-source from within.

In any event, I am emphatically not advocating some kind of mass production model for legal service. I accept that the circumstances of clients are never identical. But I do not concede that human legal practitioners are needed at all stages across the lifecycle of a legal project, even if the final output is tailored. Rather, I regard multi-sourcing and the deployment of technology as leading towards mass customization—using standard processes and systems that can meet the particular needs of clients and yet with a level of efficiency that is akin to that of mass production. A good example is document automation, as described in Chapter 4. A drafting system of the kind I describe does not simply print out a single, standard document. Instead, based on a user's answers to specific questions about his or her particular circumstances, the document generated will be one output of countless (often millions of) possible permutations. The end result is a tailored solution, delivered by an advanced system rather than by a human craftsman. That is the future of legal service.

6 | Disruptive Legal Technologies

In management theory, drawing on Clayton Christensen's influential book, *The Innovator's Dilemma*, a distinction is commonly drawn between sustaining and disruptive technologies. In broad terms, sustaining technologies are those that support and enhance the way that a business or a market currently operates. In contrast, disruptive technologies fundamentally challenge and change the functioning of a firm or a sector. An example of the former category is computerized accounting systems, which sustained and enhanced the work of those who previously laboured over paper ledgers. An illustration of the latter is digital camera technology, which famously disrupted and led in part to the eventual downfall of Kodak, whose business was based on an earlier generation of technology (chemical printing).

Two aspects of disruptive technology theory are noteworthy. First, as the Kodak example illustrates, disruptive technologies can help to unseat and bring about the demise even of market leaders. Second, in the early days of disruptive technologies, market leaders as well as their customers often dismiss the new systems as superficial and unlikely to take off. Later, however, as they gain acceptance, customers often switch quickly to services based on the new technology, whereas providers, unless they are early adopters, are often too late to recognize their real potential and never manage to regain ground.

In relation to professional workers, as Daniel Susskind and I say in *The Future of the Professions*, the term 'disruption' should be handled with care. While it is understandable that professionals may indeed feel disrupted by the types of changes discussed in these pages, we should never forget the client, the recipient of legal services. Many of the changes envisaged here should result in better or lower-cost or more convenient service for clients. It is unlikely that these beneficiaries will feel in the least disrupted. Instead, they may even feel empowered or liberated. For the buyer of legal services, this disruption is often very good news indeed. One person's disruption can be another's salvation.

There is a lesson here for lawyers that reaches beyond the issue of disruption—always give thought to the recipient of your services. When considering some kind of innovation, put yourselves in the shoes of those you are meant to be helping. What will it mean for them?

That said, given this book is largely directed at providers, it makes sense to stick with the language of disruption. One of my purposes is precisely to challenge the supply side of the legal market, that is, to contemplate the potential havoc that technology might well unleash among law firms and other legal service providers.

To be more specific, I claim there are at least 15 disruptive technologies in law (see Table 6.1). Individually, these existing and emerging systems will challenge and change the way in

TABLE 6.1 Disruptive legal technologies

document automation
relentless connectivity
electronic legal marketplace
e-learning
online legal guidance
legal open-sourcing
closed legal communities
workflow and project management
embedded legal knowledge
blockchain
online dispute resolution
document analysis
machine prediction
natural language processing
legal platforms

which certain legal services are delivered. Collectively, they will transform the entire legal landscape. In what follows, I offer very brief introductions to each of these disruptive legal technologies (and please note that there are overlaps between some of these categories).

Document Automation

These systems, as described in Chapter 4, generate relatively polished and customized versions of documents, in response to questions asked of their users. Much of the early work in this field, in the 1980s, was devoted to systems that could generate wills. Since then, the same technology has been applied in far more ambitious contexts such as the production of loan documentation for large-scale banking transactions. Document automation can therefore be used within legal businesses or made available online, and is disruptive for lawyers who charge for their time, because it enables documents to be generated in minutes whereas, in the past, they would have taken many hours to craft.

Not quite as sophisticated as full-scale automated document assembly systems are those online services that provide users with basic document templates. This was the original business of LegalZoom, a US-based company that makes legal documents available to citizens and businesses who cannot afford lawyers or wish to spend less (and here is the disruption) on their legal issues. LegalZoom and its main competitor, Rocket Lawyer, have now served many millions of customers and their brands are better known in the US than most law firms. Their service has become increasingly enabled by technology but their early success confirms that even templates, when available online with some guidance, can be useful and constitute a service that used to be the exclusive preserve of lawyers.

Relentless Connectivity

This refers to systems that together prevent lawyers from entirely disengaging from their clients and the workplace.

The technologies include handheld devices, tablets, wireless broadband access, high-definition video conferencing, instant messaging, social media, and email, all bolstered by increasing processing power and storage capacity. When these technologies combine, and the machines (of whatever kind) are switched on, which seems now to be all of the time, the 'presence' of lawyers is increasingly visible to their network of contacts. In turn, clients and colleagues will have and expect to have immediate access to lawyers. This can be disruptive for the working and social lives of lawyers. It is a sobering thought too that we seem destined to become more and not less connected, so that the disruption of relentless connectivity is likely to intensify rather than diminish.

During COVID, services like Zoom proved to be a boon and a burden—in offering immediate availability so much more usefully than by phone calls, the legal world was kept afloat. But lawyers have never felt more 'on call'. In any event, it is becoming clear that this form of interaction (inevitably much embellished) will, from now on, be fundamental to the client experience. For lawyers, who benefited from being able to charge for travel time, again there is further disruption here.

Electronic Legal Marketplace

I use this term to include online reputation systems, which allow clients to share their views, online, on the performance and levels of service of their lawyers (as customers of hotels and restaurants do); price comparison systems, which put the respective prices and rates of different legal advisers and law firms on simple websites; and online legal auctions, not unlike eBay in concept, but best suited to legal work packages that are routine and repetitive. For lawyers who used to benefit from their clients not knowing what alternatives were open to them, these technologies in isolation and together are highly disruptive. Today, these systems—social networks of a sort—are still in their early incarnations. In not too many years, they will be as pervasive as the influential printed directories that have ranked lawyers and law firms for the last 25 years or so.

e-Learning

Remarkable progress is being made on the development of online facilities to support legal learning and training. As discussed in Chapter 18, these will challenge and replace most conventional law lectures and, more broadly, will precipitate an overhaul of the traditional teaching methods of law schools generally. The techniques involved extend well beyond online lectures and webinars to the use of simulated legal practice and virtual legal learning environments. Again, the pandemic has no doubt accelerated related developments.

Beyond formal education, e-learning will also transform the way that law firms provide and integrate their training and know-how functions. We will see a move from 'just-in-case' classroom training (teaching subjects, just in case the insight provided might be needed in practice at some later stage) to 'just-in-time' learning (interactive, multimedia tools which can supply focused and tailored training on the spot).

Online Legal Guidance

These are systems that can provide legal information, legal guidance, and even legal advice on an online basis. They may or may not be subscription-based. Some can be in low-value but high-volume areas of work, such as the system originally developed by a student to help drivers challenge parking fines (this has evolved into DoNotPay), some can be in jurisdictions where access to lawyers for most people is almost non-existent, such as Barefoot Law in Uganda, and some can be in complex, commercial work, such as Allen & Overy's online legal services, which annually generate more than £22 million from over 650 subscriptions.

The threat and disruption to conventional lawyers here is clear: if clients can secure legal guidance and legal documents on an online basis, then this may come to be a low-cost competitor to lawyers whose living is made from traditional, consultative, face-to-face advisory service. And, in the terminology of Chapter 4, if robust and reliable legal help is commoditized and available at no cost to users, then it is hard to imagine, at least in some circumstances, why clients would prefer to pay good money to traditional human advisers.

Legal Open-Sourcing

Consistent with the open-source movement generally, here I envisage sustained online mass collaboration in the field of law—a movement devoted to building up large quantities of public, community-oriented legal materials, such as standard documents, checklists, and flowcharts. This is also a form of commoditization (see Chapter 4) and is disruptive for lawyers because, once again, legal content that once was chargeable as part of lawyers' service is now available for no fee. A long-running illustration of this phenomenon is the Legal Information Institute at Cornell University Law School, where, since 1992, they have been publishing law online at no charge and creating materials that help people to understand legal issues. They are sometimes referred to as a leading 'law-not-com' provider of public legal information. In health, the website <http://www. patientslikeme.com>, with over 850,000 users, is a powerful illustration of how collaborating recipients of professional service can usefully share insights and experience. By analogy, we should expect a site with some such name as 'www.legalcl ientslikeme.com' to emerge soon (take note that no one seems to own that domain name yet).

Closed Legal Communities

The idea here is for restricted groups of like-minded lawyers with common interests to come together and collaborate online in private social networks; a cross between LinkedIn and Wikipedia, but solely for the use of groups of lawyers, where users can build up bodies of collective knowledge and experience. A similar concept, again in the medical profession, has enjoyed considerable success: Sermo, an online community for doctors (no patients or pharmaceutical companies), has over

1.3 million users ('verified and credentialized physicians') from around 150 countries. The best example of this phenomenon in law was the first version of Legal OnRamp, originally described as a collaboration system for in-house counsel. Outside lawyers and third party service providers also participated. Quite quickly after launch, lawyers from over 40 countries were active, both in the general online community and in private sub-communities that could be set up. In its first iteration, it did not fly like Sermo. It came too early, in my view, but this is a space to watch.

In pursuit of the collaboration strategy described in Chapter 3, in-house lawyers are expressing serious interest in these closed communities—as platforms upon which they might share the costs of certain legal services and also as a tool to encourage and enable closer collaboration amongst their preferred law firms. For firms that are wedded to the notion of servicing their clients separately, this poses a considerable threat.

Workflow and Project Management

For high-volume, repetitive legal work, workflow systems are like automated checklists that drive a standard process from start to finish. Project management systems, on the other hand, are better suited to legal tasks and activities that are more complex, less structured, and yet still amenable to more disciplined handling than the ad hocery that is found in many law firms and in-house departments. For law firms that charge by the hour and so have historically benefited from ineffective case management and inept transaction management, workflow and project management systems represent new efficiencies and, in turn, the prospect of reduced fees.

Embedded Legal Knowledge

In years to come, in many dimensions of our social and working lives, I predict that legal rules will be deeply embedded in our systems and processes. Some people talk in this context of 'rules as code'.

Consider a car that warns its drivers and passengers that the ignition will not work until a built-in breathalysing test is used and passed. This would not require car users to know the precise details of the law and then exercise the option of applying the law. Instead, the law that prohibits driving with excessive alcohol in the bloodstream would be embedded in the car itself. Another example would be an 'intelligent' building that monitors the temperature and other environmental conditions by reference to levels established in health and safety regulations. In the event of some stipulated limit being exceeded, alarms might sound or, in emergency situations, computer screens might even be disabled. Again, this would not require people to know the law and monitor compliance. Rather, the regulations would be embedded in the building. And the building would, as it were, know about its own safety levels and make some decisions accordingly. The disruption here is that, where rules are embedded, lawyers are no longer needed to draw clients' attention to circumstances of legal significance. Likewise, self-executing contracts, possibly enabled through blockchain technology, will be able to initiate actions and automatically execute processes and provisions, without directly involving lawyers.

Blockchain

Even more so than artificial intelligence, blockchain technology gives rise to great contention. Some regard it as more fundamental than the internet itself. Others believe its likely impact is hugely overstated by its proponents. While common sense sits somewhere in the middle, most lawyers are bewildered when they seek clear explanations of what the blockchain is all about. No-one is much the wiser to learn that blockchains are a type of 'decentralized distributed ledger' or that they contain 'a series of blocks, connected together cryptographically, each with a cryptic hash'. Or, if you will forgive the pun, cryptic words to that effect. This is technobabble at its worst and for lawyers this cult-like fondness for obscure definition conceals an important reality—that blockchain technology, in broad terms, enables data and documents, first, to be stored in a way that makes it all but impossible to change or falsify and, second, to be shared amongst users with no single person or authority in control. For legal work, this is of considerable significance because this technology can enable users, efficiently and reliably, to store and share contracts, to record ownership of property and assets, to settle transactions automatically, to complete payments, to execute terms and conditions of agreements without human intervention (so-called smart contracts), and much more.

The main takeaway here is that blockchain is a disintermediating technology, which means it removes human beings for business processes and supply chains. Where these humans are lawyers, as is increasingly the case, blockchain is clearly a disruptive legal technology.

Online Dispute Resolution (ODR)

When the process of actually resolving a legal dispute, especially the formulation of the solution, is entirely or largely conducted across the Internet, then we have some form of online dispute resolution (known in the trade as ODR—see Chapter 14 for more detail and some examples). For litigators whose work is premised on the conventional, physical court-based trial process, ODR— whether in the form, for example, of online courts, e-negotiation, or e-mediation—is a challenge to the heart of their business.

Document Analysis

Lawyers spend much of their time ploughing through documents, not least in preparation for litigation. For some years now, in terms of precision and recall, properly primed systems have been able to outperform paralegals and junior lawyers when reviewing large bodies of documents and isolating those of relevance.

More recently, a new set of techniques have been adopted within law, drawing from disciplines known variously as machine learning, Big Data, and predictive analytics. Although some of the most impactful dimensions of machine learning have scarcely been deployed in legal practice (such as computers that can write programs, deep neural networks, and reinforcement learning algorithms), it is clear that these emerging systems are proving increasingly impressive, whether analysing documents sets or summarizing or extracting key provisions from contracts.

These search and machine learning capabilities are disruptive, not simply for law firms that have profited from employing human beings to wade through roomfuls of paperwork (whether on transactions, dispute-related projects, or contract reviews) but also for legal process outsourcers who currently offer similar services. No matter how low human labour costs might be, a system of this kind, once set up, will invariably be less costly. That is no doubt why young law companies that are pioneering in the field are generating great interest across the legal profession.

Machine Prediction

Another vital use of machine learning techniques is in making predictions. Increasingly sophisticated methods are being evolved to detect patterns and correlations in large quantities of data. In law, as the pioneering work of Daniel Katz showed in relation to the US Supreme Court, computational statistics (crudely, algorithms working on large bodies of data) can often yield more accurate predictions of the likely behaviour of courts than the predictions of lawyers engaged in traditional legal research and reasoning. Bear in mind the words of the inimitable Oliver Wendell Holmes: 'prophecies of what the courts will do in fact, and nothing more pretentious, are what I mean by the law'.

Much legal work involves prediction, whether of the likelihood of winning a case or negotiating a settlement, or of a deal being abandoned or completed. The data held within law firms' systems, along with publicly accessible data, will no doubt form the basis of future predictions in relation to such issues. More than this, by aggregating data sets, we will soon be able to find out what legal issues and concerns are troubling particular communities; by analysing the work of regulators, we may be able to predict compliance outcomes in entirely novel ways; and by collecting huge bodies of commercial contracts and exchanges of emails, we might gain insight into the greatest legal risks that specific sectors face. The disruption here is that crucial insights in legal practice and in legal risk management might be generated largely by algorithms operating on large bodies of data without needing to involve mainstream lawyers (unless they choose to collaborate with data scientists).

Natural Language Processing

Humans and machines, in crude terms, speak different languages—human language and code respectively. Until recently, machines have not been able to 'understand' human language, because (amongst many other reasons) it is not sufficiently structured or free of ambiguity. Scientists who work on natural language processing (NLP) try to develop systems that can handle human language with all its idiosyncrasies and illogicalities. For lawyers, there are two important projects here. The first is to enable humans to communicate (instruct and interact) directly with systems through human language and not code, while the second is to design the systems that can handle (summarize, interpret, analyse) documents written in natural human language. Earlier in this chapter, I give examples of the latter, such as the analysis of legal documents.

An important illustration of the former is known as 'question answering' (QA), which is a branch of computer science devoted to the development of systems that automatically respond to questions put by human users in everyday (natural) language. My favourite example of QA is still IBM's Watson, a system built to compete on the US TV quiz show *Jeopardy!* In 2011, on a live episode of that show, Watson very publicly beat the two best ever human competitors (see Chapter 1). In law, legal QA— a form of natural language processing—will greatly increase citizens' access to everyday law. In the spirit of Watson, this could take the form of an online service that contains vast stores of structured and unstructured legal materials (primary and secondary sources, and legal analysis) that can understand legal problems spoken to it in natural language, that can analyse and classify the fact pattern inherent in these problems, that can draw conclusions and offer legal advice, and that can even express this guidance in some computer-simulated voice (in an accent of the user's choosing, perhaps). This kind of system will disrupt not just the world of practising lawyers but also our common perception of the legal process. This is some years away yet but emerging technologies, developing exponentially, may bring this form of natural language processing to everyday law sooner than sceptics believe.

Natural language processing and machine prediction are common examples of the use of artificial intelligence in law. I return to this subject in Chapter 22.

Legal Platforms

The concept of 'legal platforms' is a relative newcomer to the world of lawtech. Until a few years ago, most legal technologists felt that the way to differentiate in the market was to develop a system or solution ahead of the market. However, most systems and solutions can be easily replicated and genuine first-mover advantage in lawtech was relatively rare. In contrast, the provider of a legal platform could potentially become the dominant provider of certain legal services, products, or solutions and in so doing disruptively exclude competing providers. We have seen this dominance in social media—think of Facebook, LinkedIn, and Twitter. We have also seen this in online trading and retail—think of Amazon and eBay. None of these businesses is the only player in their markets. But each has built huge communities of connected users, for whom there is little reason to switch to alternatives. So too in law. We can easily envisage similarly dominant platforms: perhaps an online community for in-house counsel, or a provider of industry-standard document automation, or a private-sector online dispute resolution service. If such legal platforms came to dominate, if they became the default resources, they could leave other legal providers high and dry.

7 | The Grid

In the late 1990s, at the peak of the dotcom era, practising lawyers began to take an interest in technologies other than their accounting systems, word processing, and internal email. Sensing that 'this Internet thing' might even catch on in the world of law, leaders in firms suddenly became curious about the potential of legal technology and were keen to understand what it all meant. To whet the appetite of senior lawyers and help them make sense of the field, I set myself the task at that time of creating some kind of framework that would allow me—rapidly and in plain terms—to provide a simple snapshot of the fundamental ways in which IT (as we then called it) would have impact on law firms.

The framework I came up with has now been used for more than 20 years around the world. It first appeared in the opening chapter of my book, *Transforming the Law*, published in 2000. The framework is basically a grid. It offers a rough and ready way of understanding and speaking about technology in law firms. Over the last few years, to take account of new techniques and technologies, I have refined and improved the grid and my purpose in this chapter is to introduce it in outline form.

Basic Grids

My starting point is a generic version of the grid. Forgetting lawyers and law firms for a moment, on the basic grid for any organization, as depicted in Figure 7.1, the horizontal axis represents the broad range of facilities provided under the heading of 'digital'—from basic technology (the plumbing) at one end of the spectrum to knowledge systems at the other. The vertical axis distinguishes internal use of digital facilities (below the line) from external use (above the line). As shown, this creates four fundamental uses of technology in any organization: internal use of technology, internal management of knowledge, external technology links, and provision of access

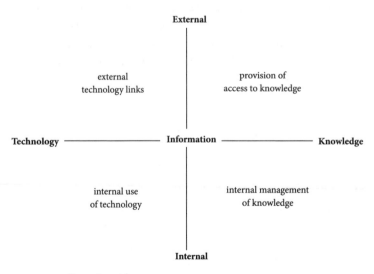

FIGURE 7.1 Generic grid.

to knowledge. The scope of each will become clearer as this story unfolds.

If we change the label at the top from 'external' to 'client', this gives us the 'law firm grid,' as laid out in Figure 7.2. This is a tool that can be used in analysing any law firm or legal provider. It shows, vitally, that there are really four basic categories of lawtech. First of all, in the bottom-left quadrant (internal and focused on technology), we have basic back-office systems that support everyday operations. Second, in the bottom-right quadrant (internal and focused on knowledge), we have systems for sharing knowledge within a firm. Third, in the top-left quadrant (external and focused on technology), we have systems that enable communication and collaboration with clients.

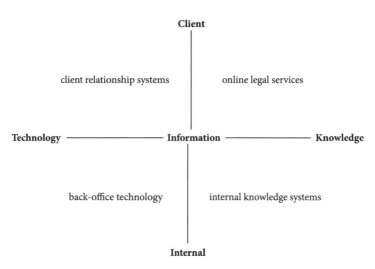

FIGURE 7.2 Law firm grid.

Finally, in the top-right quadrant (external, and focused on knowledge), we have systems that make knowledge, experience, and expertise available to clients on an online basis or embedded in clients' systems and processes (on embedded systems, see Chapter 6).

When senior lawyers are introduced to this grid, I sometimes detect an 'aha' moment. No longer is lawtech some amorphous, vague, general purpose bucket of unintelligible and costly bits of kit. Nor is it a bewildering range of systems and applications. Instead, we have but four basic categories. This simple model helps managers when strategizing, prioritizing, planning, and budgeting. More, it provides a common vocabulary so that, quite quickly, I hear lawyers say things like, 'we need to do more in the top right' or 'most of our spend is in the bottom left'.

Examples

Some examples bring the law firm grid to life. In the bottom-left, as can be seen in Figure 7.3, there are the basic systems that have been used to run law firms for many years. We used to call this 'office automation'. In the bottom right are systems that enable firms to capture, codify, and reuse their collective knowledge. In the top left are systems that support the delivery of conventional legal service by using various forms of communication technologies and by sharing data. And, in the top right are various systems that enable firms to make content, materials, documents, guidance, diagnostics, and advice available as a

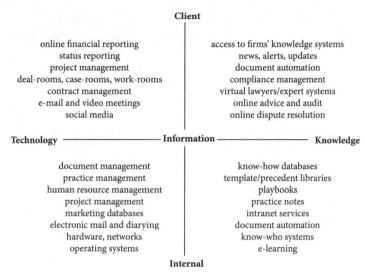

Client

online financial reporting	access to firms' knowledge systems
status reporting	news, alerts, updates
project management	document automation
deal-rooms, case-rooms, work-rooms	compliance management
contract management	virtual lawyers/expert systems
e-mail and video meetings	online advice and audit
social media	online dispute resolution

Technology ——————— Information ——————— Knowledge

document management	know-how databases
practice management	template/precedent libraries
human resource management	playbooks
project management	practice notes
marketing databases	intranet services
electronic mail and diarying	document automation
hardware, networks	know-who systems
operating systems	e-learning

Internal

FIGURE 7.3 Examples.

digital service. In my estimation and strikingly, more than 90 per cent of most law firms' spend on technology still belongs in the bottom left.

What I have said so far allows us to a create an even simpler summary of the grid and, in turn, of legal technology—see Figure 7.4. Legal technology provides as with an engine room that drives the business (bottom left), a set of tools to help firms share their knowledge (bottom right), channels through which law firms provide a digital client experience (top left), and the ability to deliver products and solutions, either through online service or by embedding content (top right).

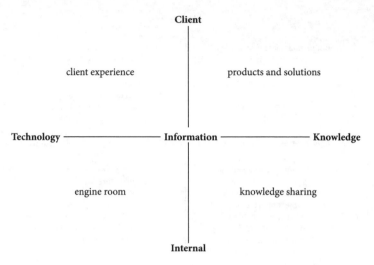

FIGURE 7.4 Summary.

Impact

Figure 7.5 shows that the four categories of lawtech bring quite different business benefits for law firms and legal providers. The bottom left is about keeping the business running, avoiding the risks of systems failure, and maintaining robust infrastructure. The bottom right focuses on making the most of the combined knowledge of the firm's lawyers—avoiding duplication of effort and reusing the best experience. The top left enhances traditional legal service, ensuring close relationships while working on jobs and between jobs. The top right represents a transformation in legal service, a move away from charging for lawyers' time to licensing a firm's knowledge: making money while you sleep.

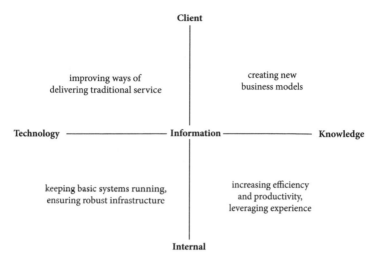

FIGURE 7.5 Business benefits.

What should clients themselves expect across the client grid? As Figure 7.6 suggests, they can reasonably ask for solid infrastructure in the bottom left (reliability and security of systems and data); competitive expertise in the bottom right (deep knowledge at sensible cost, via knowledge management or managed services); transparency and convenience (online tracking and collaborative working tools); and innovation (commoditized law).

On social media, at conferences, and in the legal press, a few technologies, plotted in rough terms in Figure 7.7, are often hailed as the most exciting current developments. It is significant that the majority of these 'hot technologies' are on or above the horizontal axis. Which means that they are systems that have

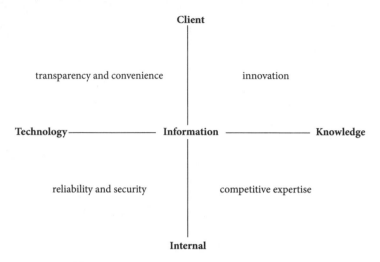

FIGURE 7.6 Client expectations.

direct impact on clients. This represents a major shift in the world of lawtech—in the first five decades of the field, until about 5 to 10 years ago, the bulk of the talked-about technologies were for use within law firms and not with clients. Although most of the overall global spend on legal technology is still, as noted, in the bottom left, there is now broad recognition that differentiation will be achieved by providers above the line.

Another impact can be plotted on the grid—that of COVID-19. It is commonly claimed that the pandemic has accelerated the uptake of technology by law firms. As I explain in Chapter 2, this is an over-simplification. In grid terms, COVID-19 in fact required law firms to invest in, and roll out, systems very largely in the top left. But the crisis has had little impact in the top right.

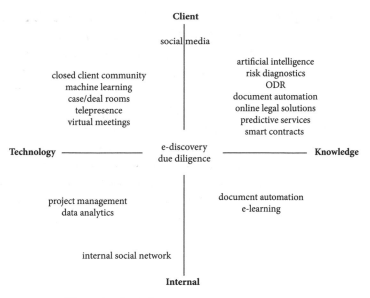

FIGURE 7.7 'Hot technologies'.

Remote-working and home-working (video meetings, email, collaboration systems) all belong in the top left, supported by the bottom left. These are not new technologies but they did prove necessary for the survival of law firms and legal businesses from 2020 to 2022. Significantly, though, these COVID response systems have not changed the business model of law firms— they continued during the pandemic to deliver the age-old advisory service. This relates to what I say in Chapter 2: during the pandemic in the legal world, we saw an acceleration of automation but a deceleration in innovation (which we would expect largely in the top right).

Client Grid

Changing the north and south labels as in Figure 7.8, an analogous client grid can be used to explain the use of technology by in-house legal departments. Here, the legal department becomes the provider and the business itself becomes the recipient of services. The bottom-left and bottom-right quadrants play host to the same types of systems as in law firms, except that in-house lawyers generally have to accept whatever business-wide systems are used in the back-office, and they tend to invest far less than law firms in their knowledge systems. In the top-left quadrant, these are the corporate relationship systems that in-house lawyers use to communicate with and advise their own

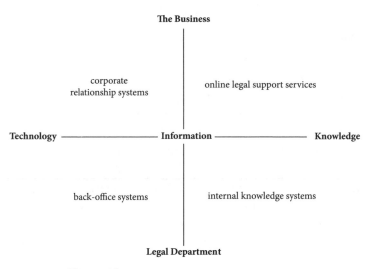

FIGURE 7.8 Client grid.

businesses. In the top-right quadrant are online legal support services, largely legal self-help tools, that in-house lawyers make available directly to their businesses. The client grid is a tool to help in-house lawyers to think about the way they use technology and plan for the future.

More generally, the emerging story of lawtech is a tale of two grids. Law firms and their clients do not work in isolation—they are connected by technology. Increasingly, when law firms make information available in their top-left quadrant—for example, in deal rooms or as online reports—there is interest amongst clients in pulling this data into their own systems, that is, into their bottom-left quadrant. Likewise, when knowledge services and online legal services are provided by law firms (their top-right quadrant), advanced clients want to have access to these collectively in one place (in their bottom-right quadrant). In this way, the law firm and client grids are coming together, as shown in Figure 7.9.

The lesson here for law firms is that most clients will soon want to gain access to their data and knowledge resources, and these clients will also want to do similarly with the firms' competitors. Competitive advantage can be secured here either by having data and knowledge resources that are superior to the competition, or by using technologies that make a given firm's data and knowledge more easily accessible. There may also be a platform opportunity here (see Chapter 6). Law firms in given fields or more generally could make available (and then dominate) industry-standard platforms for the sharing of data and knowledge.

The end game here, however, may not be with law firms or legal providers sharing data and knowledge more impressively

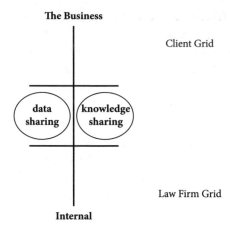

FIGURE 7.9 Convergence of grids.

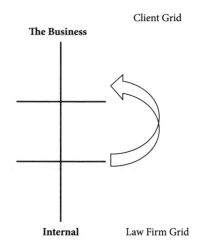

FIGURE 7.10 One end-game.

with in-house counsel than their competitors. Instead, some might seek to provide some of their legal knowledge resources (legal guidance and solutions, document automation, products, and so on) directly to end-users in businesses (see Figure 7.10). Often this will involve commoditizing everyday legal work, and so releasing in-house lawyers to focus on their specialties. Superficially, this might look as though the in-house legal function is being bypassed by external providers. However, there is a more attractive and diplomatic possibility—that leading external legal businesses work in collaboration with in-house lawyers to provide these direct-access tools.

PART TWO

The New Landscape

8 | The Future for Law Firms

One central question emerges from the first part of this book: to what extent can lawyers' work be undertaken differently—more quickly, cheaply, and efficiently, but to a higher quality—using alternative methods of working? This is a key question of the day. As noted in Chapter 3, lawyers have for many years performed routine work for which they have been over-qualified and for which, in turn, they have been over-charging. In boom times, in what has been a sellers' market, there has been little need for successful law firms to be detained by the challenge of delivering services in new and more efficient ways. Today, however, as cost pressures from clients continue to intensify, as new service providers emerge, and as new technologies are deployed, it is unwise for any firm to avoid thinking about how it should work differently.

Nonetheless, I find that many traditional law firms are not changing much, despite claims and commentary about innovation (see Chapter 21). Most are not yet adopting alternative methods of working. This is partly an issue of change management in that law firms tend to be so busy serving clients and meeting their own financial targets that they allow little time for internal reform—it is not easy to change a wheel on a moving car. It is also, in part, a structural matter, because most practices still aspire to the old textbook, broad-based pyramidic structure mentioned in Chapter 3, whereas alternative methods of sourcing call for a revision, if not rejection, of that model. And, if we are honest, there is also still reluctance amongst mainstream partners in many firms to believe that they really need to change. There is an inclination, in other words, to cling on to the old ways of working in the hope that there will soon be a solid economic recovery (from recession, Brexit, or whatever) and normal business can be resumed.

Prospects for Law Firms

However, if the analysis and predictions of Part One of this book are sound, then law firms in the coming decade and beyond will be driven relentlessly by their clients to reduce their costs. This is the heart of the more-for-less challenge (see Chapter 1). For most firms, despite their current hesitancy, I predict that this will lead eventually to the deployment and execution of alternative sourcing strategies (Chapter 5). And, in turn, we will witness the end of leverage—at best the pyramid (with partners at the top and less experienced lawyers at the base) will move from being broad based to narrow based. No longer will firms aspire to building large teams of junior lawyers as the basis of their profitability. 'To survive', in the memorable words of Theodore Levitt (in his seminal article, 'Marketing Myopia'), lawyers 'will have to plot the obsolescence of what now produces their livelihood'.

In due course, some firms may, for example, choose to strip away their junior and trainee lawyers, or stop recruiting them. They might then operate with a team of high-powered partners, each supported by, say, one associate and capable systems; and the routine work will be resourced beyond the firm. Others may elect to build their own alternative sourcing capacities, such as internal teams of paralegals, or maybe through the establishment of their own off-shored legal facilities. Still others will find opportunities for novel legal services (see Chapter 16), by creating markets that formerly did not exist or by inserting themselves in different places in legal supply chains (e.g., by becoming involved far earlier in the lifecycles of their clients' business dealings).

Although these changes will impact on all firms (large and small), some larger firms will want to argue that, for 'high-end work', notions such as commoditization, decomposing, and multi-sourcing are of little relevance. On examination, however, it transpires that this concept of 'high-end work' is something of a myth—even in the world's largest deals and disputes there are substantial components of work that can be routinized and sourced differently. And large firms that insist they only undertake bespoke work, which is a very different claim from asserting that they only do high-end work (bespoke being a subset, often small, of high-end), may find themselves at risk. They may be relegated, for instance, to the role of subcontractor to other organizations that step forward to undertake the project management of sizeable deals and disputes. At the same time, alternative providers may take up the work that these firms previously assigned to their junior lawyers.

A Global Elite?

That said, there may well be a global elite of law firms, around 20 in number today, which insist that they do not need to change much. These will be firms that continue to enjoy great commercial success. With some force, they will argue that for bet-the-ranch deals and disputes, clients will still want the services delivered, more or less, as in the past. These firms will say that for really big ticket assignments, there is only a handful of brands that will be tolerated at board level (the 'no-one ever got fired for buying IBM' principle) and that, in any event, when the future of an organization is in the balance (whether under threat or in anticipation of a great new venture) legal work is not price sensitive (the '1 million dollars here or there makes no difference in the broader scheme of things' principle). If all of these elite law firms believe this and continue to work as they have in the past, then they may well be right. And it will be hard to convince a group of millionaires that their business model is broken.

However, they should not be over-confident in their belief that, in Levitt's words, 'there is no competitive substitute for the industry's major product'. So, if a major new force (such as a 'Big 4' accounting firm) emerges, and brings a new proposition to the market—a credible brand at half the price of its competitors, for example—then this could fundamentally and irreversibly change the market, and not just for the elite firms but across the entire profession. Leaders of the elite firms should suspend their likely incredulity at this scenario, if only because major clients, as never before, are commonly saying that they are looking for alternatives to the traditional ways of some of the great firms whom many regard as too costly and sometimes too arrogant. Some clients, for example, are decomposing corporate work and passing the routine elements (such as due diligence) to lower cost, regional firms or to legal process outsourcing or managed service providers, and inviting the elite firms only to focus on the most challenging elements of the work. This can strike at the heart of the profit model of large firms, who rely on their junior lawyers doing these routine tasks.

Elite firms should also beware of one of their number breaking rank—we saw an example of this in 2016, when Allen & Overy, in collaboration with Deloitte, launched an online system, MarginMatrix, to help banks cope with the documentation challenges arising from the new rules governing the global derivatives markets. This system pulled the rug from under the feet of many elite firms who had expected to tackle this regulatory burden for their clients in a bespoke manner.

Meanwhile, the other 'magic circle' law firms based in London (Clifford Chance, Linklaters, Freshfields, and Slaughter and May) have also been investing in new labour models and client-facing technology. In the US too, leading firms like Cleary Gottlieb and Wilson Sonsini have been stretching the conventional boundaries. Whether these or other major firms will make more radical changes remains to be seen.

As for medium-sized firms, to survive and thrive I suspect most will need to merge and seek external investment to enable the changes from their current approach to a new, sustainable, longer-term business model. There is a window of opportunity here—they should recognize that clients' dissatisfaction with some of the leading firms throws up an unprecedented opportunity to be recognized as credible alternatives. To do this, they must find ways of building their reputations, brands, and capabilities.

I believe there will be a market for many years yet for small to medium-sized firms with demonstrable, niche expertise. General Counsel of even the largest of organizations often indicate that they welcome deep expertise and personal service even if offered from modest-sized firms. Usually, it is the talents of a particular lawyer rather than of a particular firm that is the attraction here.

As for much smaller firms with very few partners, aside from those which also offer a genuinely specialist or personal service that some market is prepared to pay for, I find it hard to imagine how these legal businesses will survive in the long run unless they change fundamentally. The threats will come from various directions, not least from online legal services providers. On the High (Main) Street, in liberalized legal regimes, retailers and banks will also compete with sole practitioners and small firms for everyday legal services (such as conveyancing, probate, and personal injury work). But it is likely that these alternative business structures, fuelled by external investment and driven by experienced business managers, will standardize, systematize, and externalize legal services (see Chapter 4) and bring cost savings, efficiencies, and experience that traditional, small law firms will find impossible to match. This will be the end of lawyers who practise in the manner of a cottage industry. I do not see much of a future as we move through the 2020s for most traditional small firms in liberalized regimes.

Of all the observations I made in the first edition of this book, my pessimism about small firms seemed to cause greatest offence.

I should say immediately that I have nothing against small firms. I just cannot see how these corner shops will manage to keep afloat in the face of the metaphorical supermarkets, whether larger businesses or online services.

Another way to look at this is for partners in small firms to ask themselves this question—in the 2020s, what unique value can we bring as a small legal business? Here are a few answers that pass muster: our clients have told us unequivocally that they do not want us to change; our community requires a wide range of legal services and there are no obvious competitors to our traditional offering; we are acknowledged specialists and, although small, we are as expert as anyone; although we are lawyers, our clients come to us because we are their general business advisers; because we are smaller, we can offer a high-quality service at significantly lower cost than any other legal businesses; our clients come to us because they are prepared to pay extra for a high touch, face-to-face service. Some small firms may feel they can genuinely provide such answers, but not many.

Trial Lawyers and Barristers

Another group of legal specialists who often maintain that they will be unaffected by economic forces, liberalization, and technology are barristers in England and trial lawyers in law firms around the world.

It is true that much of the work of the oral advocate is highly bespoke in nature and it is not at all obvious how the efforts and expertise of the courtroom lawyer might be standardized or computerized. Indeed, oral advocacy at its finest is probably the quintessential bespoke legal service. I have little doubt, for the foreseeable future, that very high-value and very complex legal issues will continue to be argued before conventional courts in the traditional manner. When there is a life-threatening dispute, clients will continue to secure the talents of the finest legal gladiators who will fight on their behalf. However, it is less clear that instructing barristers or trial lawyers for lower value or less complex disagreements will continue to be regarded as commercially justifiable. Quite apart from a likely shift towards mediation, collaborative lawyering, and other forms of alternative dispute resolution, emerging techniques of dispute containment and dispute avoidance (see Chapter 12) are likely to reduce the number of cases that find final closure in courts of law or even on the steps of the courthouse. Moreover, courtroom appearances themselves will diminish in number with greater uptake of virtual hearings, while online courts and online dispute resolution (ODR) will no doubt lead to the displacement of many conventional litigators. The future for truly exceptional senior trial lawyers and barristers looks rosy for some time yet, therefore, but junior civil trial lawyers may need to rethink their prospects and prepare for daily work in virtual hearings and online courts if they wish to prosper in the coming decade.

In England and Scotland, those barristers and advocates whose practices are devoted to the writing of opinions on complex areas of law will be less affected by the changes anticipated in this book than most other areas of legal practice, because there is no obvious alternative source for this genuinely bespoke work.

Questions Asked by New Partners

In the context of great change and disruption within law firms, I notice that recently appointed partners of law firms are currently disconcerted and nervous. In the past, I have found young partners to be an interesting breed. Often in their mid-thirties, these bright young lawyers have great energy and considerable experience but have tended to operate on the expectation that the firms of which they are now part-owners are likely to function and profit in the future much as they have in the past. They have tended to be a confident bunch, pumped up by recent admittance into partnership by their seniors, and satisfied that the effort they have expended has been justified; although they have often been unnerved to discover that becoming a partner is of itself a new beginning and to find themselves on the bottom rung of yet another ladder.

In the last few years, however, I have noted that junior partners are less confident in their position and worry very deeply about the future of the firms they have joined. When I addressed law firms' induction courses for new partners between 1996 and 2006, I felt them dismissive of my seemingly outlandish ideas. Most preferred to look at their (then) BlackBerry devices or draft documents during my presentations. This has changed radically. Today junior partners are all ears, consistently asking me the same set of questions, and are anxious to hear my views. Here are these questions, along with the replies I usually give.

Is our firm viable and sustainable?

The changes I anticipate in Part One of this book are already taking hold. If firms do not embrace alternative sourcing strategies, in particular, I doubt the majority will be viable or sustainable in the long term.

Is the business model broken?

Insofar as this refers to the broad-based pyramid with the partner at the top and junior lawyers undertaking routine work at the bottom, then, again, I think this model will indeed be broken in the long term. Leverage will be replaced by alternative sourcing.

Have the glory years passed?

For many firms, I believe their peak was around 2006. This is not simply a matter of profitability and turnover, for many have greatly improved these since, but a question of the ease with which work is won, the level of fees that clients can be charged without challenge, and the amount of human effort expended. Some elite firms and entrepreneurial firms will go on to enjoy yet greater times, but for many firms, unless they change radically, the glory years have indeed passed.

Are our fixed costs too high?

In the coming years, firms will need to revisit their property strategies, because high rentals in expensive cities will be a costly indulgence in an increasingly networked world with pervasive use of video meetings; and the labour costs of large numbers of junior lawyers will also be excessive, largely because of the availability of alternative sourcing in lower-cost regions or countries.

What are we inheriting?

Most junior partners are inheriting outdated, outmoded, low-tech businesses that will soon not be fit for purpose. This means not that there is a lack of great intelligence or talent within the businesses but that the way in which this talent is taken to market is no longer competitively arranged and priced.

Do senior partners care about the long term?

This is a vital question. Regrettably, most law firm leaders that I meet have only a few years left to serve and hope they can hold out until retirement before much that I predict engulfs them. Operating as managers rather than leaders, they are more focused on short-term profitability than long-term strategic health. For junior partners this is tragic because any major reinvention and re-engineering of law firms has to be driven from the top. I find a contrast here with the large accounting firms, where senior partners seem more concerned about the prospects of their junior partners. Their philosophy—to regard themselves as temporary custodians of long-term and enduring institutions rather than short-term investors who want to bail out when the price is right—is one that could fruitfully be assumed by more equity partners in law firms. In other words, it is time for senior partners to think more deeply about a more generous legacy for their successors.

Time for Leadership

Managers and leaders have different job descriptions in law firms. Managers tend to focus on the short term, ensuring that clients are served, work is won, the figures are delivered, and the teams are motivated. The best leaders, in contrast, do keep an eye on the short term but at the same time are obsessed with the long-term strategic health of their organizations. Their preoccupations should be with understanding changes in the market, building enduring relationships, enhancing the brand, and reviewing and developing strategy. Historically, from the 1980s through to the recession, most law firms did not need to worry about leadership. The game plan was simple—most practices were invited, year on year, to bring in a bit more work and trim their expenses. This was fair weather management, a light hand on the tiller. In times of immense change, however, much stronger direction is needed. To endure and prosper, firm leadership is required.

The finest leaders in tomorrow's law firms are unlikely to proceed in the manner of their predecessors, which was by building consensus. When change is rapid and pervasive, there will not be time to convince the doubters and laggards.

In the end, most lawyers are convinced by evidence and not argument. For every argument, a bright lawyer can concoct a clever counter-argument. In contrast, it is harder to gainsay evidence such as news that some client has enthused about a particular system or innovation. So leaders need to provide evidence in support of change rather than mere rhetoric. The problem here is that if this evidence comes from beyond the business, then the firm is, by definition, already a follower. Therefore, leaders will need to generate evidence from within— from pilots, experiments, and testing ideas out on sympathetic clients. Law firm leaders will need some leeway to undertake ongoing research and development without needing to seek the approval of the majority of the partnership. It may sound less partnerial and collegiate but it is certainly more business-like.

9 | The Shifting Role of In-House Lawyers

Some of the most fulfilled lawyers that I meet work in-house. This means they belong to legal departments that sit within large organizations. Some of these departments—for example, in large financial institutions—can be very large, with as many as 2,000 or so lawyers. A career as an in-house lawyer attracts those who want to be at the heart of the organizations that they advise. If you work in a law firm, you are at one remove from the business of your clients. If you work in-house, you are part of the business. Most law students, when contemplating their future careers, tend to imagine themselves working in law firms (other than in the US, where many aspire to being government lawyers). In many law schools, there is little formal discussion of the role of in-house lawyers, which is bizarre because these clients are likely to have an enormous influence on the future of legal services.

Legal Risk Management

Most General Counsel (GCs), the leaders of in-house legal teams, tell me that their principal job should be that of managing risk;

that 'legal risk management' should be the core competence and service of in-house lawyers. They often contrast this with what they actually do, which is fight fires—in-house lawyers are faced, on a daily basis, with a barrage of requests, problems, and questions from across their organizations. And they usually feel they have to respond helpfully. In reality, while some of these inquiries merit serious legal attention, others assuredly do not. The hope of most GCs is that they can organize themselves to become more selective; that they can move from being excessively reactive to being proactive. In other words, their job should be to anticipate problems before they arise. The focus should be on avoiding disputes and problems rather than resolving them.

Legal risk can be managed in many ways, but the emphasis is usually on preventing non-lawyers in businesses from inadvertently exposing their organizations to some kind of liability (such as might flow from a breach of some regulation or of an agreement). This control of risk can be achieved, for example, by increasing legal awareness, by introducing protocols or procedures, by using standard documents, or by involving lawyers more directly in the affairs of organizations. Legal risk management can also involve the conduct of audits, risk reviews, and health checks to assess, for instance, an organization's processes for managing regulatory compliance or its preparedness for litigation. There is little question that tomorrow's in-house lawyers will become increasingly systematic and rigorous in their management of risk and will require sophisticated systems and techniques to help them (the most exciting opportunity here is the use of machine learning techniques for risk diagnosis). Strikingly, very few law firms have yet recognized the commercial opportunities here.

Another risk-related trend will be towards the greater sharing of risk between in-house lawyers and law firms. If deals and disputes do not conclude satisfactorily, some GCs believe that the law firms involved should suffer some of the downside, by reducing their fees. With some justification, law firms retort that this should cut both ways, so that the successful conclusion of a legal project should surely then result in an uplift in fees. No doubt, these debates on fees and risk-sharing will intensify in years to come, as economic pressures increase. New ways of allocating risks will evolve in attempts to incentivize law firms in different ways. One arresting example of this is when in-house lawyers pay law firms bonuses if they help them to avoid litigation.

Knowledge Management

The use of standard documents, as said, is a well-established technique for reducing legal risk: non-lawyers and lawyers alike are required to use (and only then with permission and sometimes supervision) fixed-form agreements that have been carefully crafted in anticipation of well-known legal problems and pitfalls. Business people can be constrained in their negotiations by imposing the use of agreements with terms and conditions that cannot be altered without sign-off from lawyers.

The actual preparation of these standard documents belongs to the world of legal knowledge management. This is the process of capturing, nurturing, and sharing the collective know-how and expertise of a group of lawyers. The motive here is to avoid duplication of effort and to build an institutional memory that is superior to the recall of any individuals, no matter how talented. Knowledge management is one of the central jobs of professional support lawyers, a key group of legal specialists who work in major law firms, especially in the UK.

Significantly, in-house legal departments rarely employ knowledge managers and professional support lawyers. There is a paradox and inconsistency here. It would clearly be in the interests of in-house lawyers to secure the efficiencies that knowledge management would bring. In contrast, for law firms that charge by the hour, the incentive to become more efficient through knowledge recycling is less than immediately obvious. Why, then, do in-house lawyers generally hold back from recruiting knowledge managers whereas major law firms have invested heavily? For in-house lawyers, the deterrent seems to be the expense of employing professional support lawyers—it is difficult, I am told, to make the business case to chief finance officers for employing lawyers who do not advise directly on everyday legal work, or on disputes or deals. As for law firms, they know that their clients (in the UK, if less so in the US and Canada) expect their external advisers to have substantial bodies of templates and precedents, and knowledge managers are the people who specialize in maintaining this kind of know-how. In summary, most in-house lawyers like the idea of knowledge management but would prefer law firms to pay for it.

This will change. In due course, in-house lawyers will recognize and be able to quantify the benefits that professional support lawyers can bring and will manage to convince their boards that it makes sense to invest in people who will bring savings through technology-enabled legal knowledge and process sharing (within legal departments and between organizations too).

Expecting More from Law Firms

Moving away from risk and knowledge management, how will clients select law firms in the future? It is often assumed that what differentiates one lawyer or law firm from another is their substantive expertise; that clients will gravitate towards lawyers who seem to know more or appear more deeply expert. However, clients often say that there is little to choose between many good lawyers and good law firms, that they are equally and impressively familiar with black-letter law and market practice. What frequently distinguishes law firms, particularly when the work is genuinely bespoke, are the personal relationships that lawyers have with those they advise. (When the work is routine, the interpersonal dimension is of less importance.)

To run a successful legal business in the future, therefore, it will not be sufficient for lawyers to be in possession of fine legal minds. Tomorrow's lawyers will need to acquire various softer skills if they are to win new clients and keep them happy. In-house lawyers of the future will not only be more demanding on costs, inclusion and diversity, and climate issues; they also will be more discerning about the interpersonal relationships they choose to cultivate with lawyers in external firms. This will place pressures on law firms to make the most of face-to-face interactions and to use social networking systems and video meetings to maintain regular contact.

Clients already respond favourably, for example, to law firms that express ongoing, and even passionate, interest in them. They like to feel that the firms to which they pay substantial fees are bearing them in mind and have their interests at heart, even when not working together on a particular job. They appreciate those law firms that have clearly devoted their own time to thinking specifically about them and their business and their industry. Clients like to hear, for instance, about a deal that has been done that may be relevant to them. They appreciate periodic briefings on the trends and developments that may have a direct impact on them. Maintaining this sort of rolling contact does not come naturally to many lawyers and is often trumped by pieces of chargeable work for other clients. This is regrettable because this kind of regular interaction is increasingly vital for the long-term relationships that clients are now deeming important.

A related issue to which young lawyers should be sensitive is the need for law firms to empathize with their clients. GCs often observe that their external law firms do not understand their clients, that they have little insight into the daily dynamics and operations of their clients' businesses. It is not that the law firms fail, for instance, to read their clients' annual reports (although some do fall at this hurdle) or that they are ignorant of fundamentals of the sector in which their clients trade. Instead there is a wider worry: that law firms do not take sufficient time to immerse themselves in their clients' environments and get a feel for what it is actually like to work in their businesses. For example, it has been suggested to me that most firms do not grasp, in any given client, the tolerance and appetite for risk, the amount of administration and bureaucracy, the significance and extent and tone of internal communication, and, vitally, the broader strategic and business context of the deals and disputes upon which they advise.

In short, tomorrow's lawyers will need to be more in tune with tomorrow's clients. In contrast, when meeting with their clients today, many partners of law firms are said to broadcast and pontificate instead of listening to what is actually on the minds of those they are serving. In other words, many law firms lack empathy. They fail to put themselves in their clients' shoes and see the business from the clients' perspective. It is often claimed that, because they do not pause to listen, firms cannot distinguish between those occasions when a client wants quick, rough-and-ready guidance as opposed to detailed and exhaustive legal analysis. This lack of empathy and inability to listen could be deeply prejudicial to long-term relationships between firms and clients in the future.

The More-for-Less Challenge

Although legal risk management and knowledge management will be key strategic issues for tomorrow's in-house lawyers and the quality and tone of their relationships with firms will be an important operational concern, the dominant management preoccupation of most GCs today is meeting the more-for-less challenge (see Chapter 1). For the foreseeable future, this is one thing that keeps most GCs awake at night. How can they deliver more legal service to their businesses at less cost?

The low-hanging fruit here is the possibility of driving down the fees of external lawyers. But there is a primal and fundamental tension here because clients and lawyers have very different objectives. When a client phones a law firm and intimates that their business has a problem, it is an unusually virtuous partner who will not hope, deep down, that it is a big problem. For any piece of legal work, the client will invariably pray that their legal requirements are routine and can be disposed of quickly and painlessly, while a law firm will generally hanker after more challenging instructions that might occupy a team with complex work for quite some time.

There are other related tensions arising from the still-dominant practice of hourly billing. Most clients do not want to buy the time of experts. They want results, solutions, outcomes, and practical commercial guidance. They also want certainty and predictability of costs, and not the open-ended commitment of the blank cheque that hourly billing often entails. Generally, hourly billing does not incentivize law firms to give clients what they would actually like. Consequently, we will see in the coming decade, as noted in relation to risk management, more sophisticated mechanisms for aligning the incentives of law firms and their clients.

These mechanisms will not be crude and ineffectual alternatives to hourly billing. In Chapter 3, I explain why these generally disappoint. Instead, in-house lawyers will come to the view, as discussed in that chapter, that the cost savings they need cannot be secured simply by pricing differently. Rather, the challenge is to work differently. Increasingly, in-house counsel are arriving at this conclusion and so are now wrestling with various alternative ways of sourcing legal services.

The underpinning thinking here bears repetition. Historically, legal work has been undertaken either by clients themselves or by their outside law firms. The problem with this is that it is proving too costly for routine and repetitive legal tasks to be discharged within firms and legal departments. And so, different approaches to sourcing such work have gained some traction: outsourcing to third party providers in low-cost countries; off-shoring legal work to locations where businesses have already transferred other functions, such as call centres; encouraging law firms to subcontract to practices in less costly regions; or using contract lawyers who charge about half the price of traditional law firms. These are all instances of what I call, in Chapter 3, the 'efficiency strategy'—cutting the costs of legal service. In the 2020s, technology-based solutions will become commonplace for those who are pursuing this strategy.

Yet another possibility is co-sourcing, which can involve a group of in-house departments coming together and sharing the cost of some common legal service, perhaps by setting up shared services centres. This is an example of the 'collaboration strategy' outlined in Chapter 3, where I mention the ways in which banks and local authorities are already cooperating.

There is no doubt that the in-house community is becoming steadily more interested in these and many other new ways of sourcing legal work.

The Collaborative Spirit

A different form of cooperation is also emerging—some in-house lawyers are keen to engender a collaborative spirit amongst their external law firms. They speak of their primary law firms as their 'extended family'. The intention here is that firms trust rather than compete with one another, and that their collective energies are directed at supporting the client instead of jockeying for position for the next tranche of work. The result should be a more productive, efficient, and civilized group of lawyers. On this view, the legal capability of an organization is the combination of the in-house function and its external firms. The lawyers from the firms are expected to work together as a family—not one that is dysfunctional and constantly bickering but one that shares and focuses relentlessly on a larger common purpose: the interests of clients.

This approach to managing external law firms is not yet common. Indeed some GCs are sceptical about inter-firm cooperation. Many banks seem to fall into this camp. They maintain that it is plainly unrealistic to expect their principal external firms to collaborate. Hard-nosed lawyers want a market and not a social club or a family outing. Some in-house lawyers therefore actively encourage their firms to compete strenuously with one another. On this more combative approach, firms are frequently invited to bid against one another, and to demonstrate their supremacy—that they are better, less costly, more efficient, or more innovative than the rest.

Although there are no right answers here, I have seen both schools in action (within and beyond the financial services sector) and predict that the collaboration camp will eventually win out. This approach holds obvious attractions: duplication of effort can be avoided; asymmetries can be eliminated; energies are more efficiently channelled towards the clients; and working relationships are more amicable. It simply makes sense, for example, from the clients' point of view, for their external firms to coordinate in the provision of training services. Interesting opportunities emerge, such as assembling 'dream teams', made up of the best lawyers, hand-selected from across various firms and purpose-built for particular deals and disputes. The challenge for those who favour family over feud is to put the incentives in the right place, so that law firms genuinely want to cooperate rather than compete. Half the battle here is for the client to ensure a more or less steady flow of work for firms who are family members. It will make sense on this collaborative approach for participants to embrace social networking technologies. These will bring firms under the one virtual roof and encourage and enable them to work in virtual groups. This could be done using generic services such as LinkedIn or legal tools like the original version of Legal OnRamp. As in so many other areas of legal practice, the future for in-house lawyers will be digital.

Strategy for GCs

In practical terms, how are GCs preparing for the future and, in particular, addressing the more-for-less challenge? I cannot answer that question across the board but I have found that four broad strategies are in play, each differing in its scope and ambition. The first strategy has been for GCs to concentrate largely on external law firms and to drive their prices down. This is the preferred method of GCs who pass much of their legal work to these external law firms. The second approach, better suited to large in-house departments, has been to focus instead on reshaping the in-house departments. The third has been simultaneously to review internal and external capabilities and seek to streamline both. The fourth tack has been the most ambitious—to start with a blank sheet of paper, to forget the current resources (in-house and outside), and instead to undertake a comprehensive legal needs analysis for the business. Once this analysis has been completed, the task then has been to identify dispassionately how best to resource the full set of needs; drawing not just on conventional lawyers but on alternative legal providers too. This final strategy, in my view, is the one that will deliver the most cost-effective and responsive legal services for large businesses in the future and, in due course, will be the preferred approach of all competent in-house functions.

One related development during the last decade must be noted—the appointment of chief operating officers (COOs) or directors of operations in in-house legal departments. These individuals are charged with the task, broadly speaking, of running the department like a business. Many are focused on strategy, alternative sourcing, more effective procurement, and technology, while their GCs are freed to do what they do best—acting as counsel to the generals. The COOs themselves are also collaborating: the most shining example here is CLOC, the Corporate Legal Operations Consortium, a community of COOs set up in 2014. CLOC describe themselves as a global community of experts that is redefining the business of law. More than 2,000 people attended their major 2022 event. Their influence (and combined purchasing power) is growing and considerable. I have little doubt they will play a central role in setting new standards for the legal industry.

The Power and Responsibility of In-House Lawyers

I often find, somewhat surprisingly, that in-house lawyers betray a lack of self-confidence when contemplating the future.

Frequently they ask me if I expect law firms to revert to their old ways of working in rosier economic times and when the pandemic is firmly in the rear-view mirror. I invariably respond that it is almost entirely up to them, as the customers, to shape the answer to that query. If in-house lawyers do not want the reinstatement of bad past habits, they must send that message very clearly to their external advisers. Major clients can be assured that such a message cannot be ignored, even by the most flush elite law firms.

Most in-house lawyers will concede, in principle, that change is necessary and that they should run a tighter ship and drive a harder bargain with their suppliers, but most also claim that they do not seem to have the time, energy, or competence to introduce efficiency or collaboration solutions. When I probe more deeply, it transpires that many GCs would prefer off-the-shelf answers developed by law firms. However, and this is something of a vicious circle, there is, as noted, little incentive for law firms themselves to support either the efficiency or collaboration strategies. Why should law firms destabilize their current businesses with potentially disruptive innovations when their clients often seem indifferent and their competitors are not undercutting them?

In-house lawyers must also remember that they are likely themselves to come under the microscope within their own organizations. It will not be plausible for them simply to complain ad infinitum about law firms' unwillingness to change. As it becomes widely known, for instance, that it is possible to source legal work in different ways, chief executives, chief finance officers, and boards will inevitably ask their GCs whether their departments are adapting and exploiting the opportunities afforded by these new ways of working. To help focus in-house lawyers' minds, I express this likely demand in terms of what I call the 'shareholder test':

When a costed proposal for the conduct of a deal or dispute is being considered, would a commercially astute shareholder, who was familiar with the growing number of alternative ways of sourcing legal work, consider what is contemplated as representing value for money?

If in-house lawyers allow law firms to continue with their age-old billing and working practices, they plainly fail the shareholder test. Soon GCs (and their COOs) will have little choice but to overhaul their departments and working practices: the more-for-less pressure will build to an almost intolerable level and they will have to recalibrate if not re-engineer the way they work internally and how they source external legal services.

In-house lawyers will flourish only if they can add relevant value that cannot be delivered by competing sources of legal service. The genuinely expert and trusted in-house legal adviser, who lives and breathes the business, should always be an invaluable resource, but unless GCs are also prepared to drive the efficiency and collaboration strategies within their own departments and across law firms as well as other providers that serve them, then their future is far from clear. I sense now that the in-house legal department platform is starting to burn. Now is the time to meet the challenge of change.

They should remember (although many do not seem fully to grasp this) that they have immense purchasing power. I struggle to understand why GCs have not driven external law firms much harder. The world's leading 100 law firms are sustained very largely by the world's top 1,000 businesses. If and when GCs become radically more demanding, they will have it within their power to urge a reshaping of this top echelon of firms and, in turn, redefine the entire legal marketplace.

10 | Lawtech Start-ups

Fundamental change in the legal world is unlikely to be driven over the next few years by today's incumbent providers. So long as leading law firms are trading profitably, it is generally not in their commercial interests to stop or divert their gravy train. Indeed, as Daniel Susskind and I report in the updated edition of our book, *The Future of the Professions*, we have yet to find any prominent professional firm that has significantly self-disrupted. In relation to technology, the leading legal players are of course spending liberally on their systems but the bulk of this investment is invariably to enhance and support their current business model rather than displace it.

In contrast, the accounting giants and other alternative legal providers (Chapter 17) are more likely than law firms to seek to effect significant change through technology. They are disruptors and are building at least some of their services from scratch and doing so largely on digital foundations.

More dramatic still, however, could be the disruptive impact of the burgeoning community of lawtech start-ups. In numerous other sectors—retail, leisure, transport, and entertainment, for example—transformation has been precipitated not by existing players but by remarkably motivated, small, new outfits that have dared to upend the status quo. Likewise, in law, I cautiously expect that a handful of exceptional start-ups will be prime movers in instigating pervasive and radical change.

The Start-up Landscape

I use the term 'start-up' loosely here. In broad terms, I am referring to lawtech businesses that have been founded or have mushroomed in the last decade or so. This will therefore include 'scale-ups', 'early-stage' businesses, as well as some more mature companies.

I know that venture capitalists and others have much richer vocabularies here for labelling tech companies, based on factors such as their type and timing of financing, the style and spirit of their management, and the phase and polish of their product development. My focus is different. I am not concerned, for current purposes, with the business side of these start-ups. My fascination is with their impact on the world of law, and so with the people who are thinking fundamentally afresh and are introducing and executing novel ideas in the legal market. The start-ups that fire my imagination are the ones that bear little relation to the traditional suppliers (law firms, courts, legal publishers, law database providers, and basic back-office technology vendors). I know that many or most lawtech start-ups will not survive. But the small cohort that do prosper could well redefine the legal ecosystem. This is hard for traditional lawyers to accept but that is the very nature of the kind of change I predict—mainstream providers, even the very best, find it difficult to imagine an alternative to today's prevailing orthodoxy.

How many lawtech start-ups are there? I have not been able to find a single, scholarly global study of these businesses. A useful starting point is the well-structured database assembled a few years ago by the CodeX team at Stanford University. Taking this together with other available online resources and having conducted an informal survey myself (with colleagues and contacts in around 15 countries), I estimate that there are currently between 3,000 and 4,000 lawtech start-ups around the world. When I published the first edition of this book in 2012, there were only a few hundred. At that time, the launch of a new one was a big deal. In contrast, it seems like an almost daily occurrence for me to learn about a new legal venture, product, or service (countless incoming emails keep me up to date; as does *Artificial Lawyer*, the news and analysis service).

These lawtech start-ups are now attracting considerable investment. In mid-2022, the *Financial Times* reported that lawtech companies 'had topped $1bn in venture capital investments in the nine months to September 2021, beating the previous high of $989m in 2019'. The scale of investment has increased by an order of magnitude since 2017. There are now a few unicorns (valued at more than $1 billion) in the sector, such as Ironclad, LegalZoom, and Clio. Analysts expect more to follow.

Alerted no doubt by these figures, some law firms have recently taken a more active interest in start-ups, and not just by monitoring their progress. Some act as their legal advisers, others mentor founders, yet others invest, and a few act as informal incubators and accelerators. A small number of law firms have gone further. In 2017, for example, Allen & Overy launched Fuse, its 'tech innovation hub'. This lively space in the firm's London headquarters is shortly to host its sixth annual cohort of start-ups working in lawtech, regtech, and dealtech. In the same year, and also in London, the law firm Mishcon de Reya, announced MDR Labs as a lawtech incubator dedicated to 'early-stage companies who are trying to transform the legal industry'.

Other leading professional firms have followed: for example, Slaughter and May has a lawtech programme called Collaborate, which brings developers and clients together, and in Spain, PwC and Microsoft joined forces recently to create Springboard, which is a tax technology and lawtech accelerator for European companies.

In the academic world, a lawtech incubator sits at the heart of the Legal Innovation Zone at Toronto Metropolitan University (formerly Ryerson University), set up back in mid-2015. Governments are also taking an active interest in lawtech start-ups. In the UK, for instance, LawtechUK is a government-funded initiative that promotes the greater uptake of legal technology. Its Sandbox Programme connects start-ups with potential users and collaborators. It also offers help with regulation and in securing access to data sources. Likewise, in Singapore, its government-backed FLIP (Future Law Innovation Programme) helps incubate and accelerate lawtech businesses.

Scope

What new systems and services are these lawtech start-ups bringing to the market? In no particular order, the application areas that I see as dominating are regulation and compliance, online legal research, practice management, e-discovery, document analysis, contract life-cycle management, contract drafting, e-learning, workflow, legal self-service, legal diagnostics, case management, and transaction management. The main enabling technologies are artificial intelligence, machine learning, natural language processing, alongside blockchain and, more prosaically, collaboration and workflow systems.

Dipping down further, the most common focus is on contracts. The terminology here is wide-ranging: digital, automated, and electronic agreements; contracts that are in some sense 'smart', 'smarter', 'self-executing', and even 'intelligent'—new ways to splice, dice, and price contracts. In practice, this means systems variously for negotiating, co-authoring, approving, drafting, generating, authenticating, signing, timestamping, registering, storing, hosting, sharing, searching, navigating, analysing, reviewing, querying, standardizing, templating, comparing, extracting, predicting, summarizing, self-servicing, commercializing, creating workflows, managing lifecycles, and more. The hype, the hyperbole, and even the grammar are heart-warming or exasperating, depending on your angle. On any view, though, all of this is some distance from quill and parchment.

Impact

To convey the ambition and spirit of lawtech start-ups, I used to say that each is trying to do to law what Amazon has done to bookselling. I now regard this as inaccurate. On more detailed analysis, it transpires that only a fraction—in my estimation about one-quarter—of these start-ups are seeking to bring upheaval and transformation in the world of law. In the language of Clayton Christensen (Chapter 6), there are still many more 'sustaining' than 'disruptive' start-ups. By this I mean that most of the lawtech start-ups are working on systems and services that, in broad terms, enhance rather than replace traditional legal service. The target customers of most of today's lawtech start-ups are law firms and, generally, larger rather than smaller practices. This is entirely understandable. As I note in Chapter 1, the lion's share of spend on lawtech comes from law firms, and so it makes good business sense for founders to develop systems that are likely to land well in a market with well-established patterns of expenditure.

To pick a few clear-cut examples—start-ups that offer law firms new tools for, say, time recording, conflict search, cash flow improvement, and customer relationship management are not redefining the legal market. These tools may be invaluable for their users and may enjoy great commercial success but they are not Amazon-in-law. Consider also the many systems from start-ups that focus on contracts, as discussed above. Most of these are directed at law firms, with far fewer for exclusive use by in-house legal departments, smaller businesses, and individuals. Again, I can see the commercial good sense here, but when law firms use systems that help them negotiate and draft agreements more efficiently, this is a case study in *automation* rather than disruptive *innovation* (using these terms in the specific senses discussed in Chapter 1. By contrast, some start-ups are delivering systems that enable businesses to handle their own preparation, flow, and execution of contracts, with no or minimal involvement of lawyers. These are disruptive. They do not perpetuate age-old practices. They offer a glimpse of a new world, a less-lawyered world, but one in which businesses can manage their own legal risks.

Does it matter that lawtech start-ups are automating more than innovating? For founders and investors, it may not, although I expect successful disruptive innovators will stand a better chance of becoming tomorrow's lawtech unicorns than those start-ups that graft technology onto traditional legal businesses. The disruption route may be riskier but it is probably the most promising route in the 2020s for eye-watering returns.

For me and, I suspect, for many readers, focusing on innovation more than automation matters for another reason. Our legal and court systems are creaking. Some would say they are broken—unaffordable and inaccessible. And the hope that I and others harbour is that the best of the lawtech start-ups will give us the tools to help us fix these problems. I have in mind two game-changing categories of start-up. First, there will be those that genuinely do disrupt and obviate the need for the traditional supply chain of traditional lawyers. In the spirit of Amazon, Uber, and Airbnb, we await the great coming of revolutionized legal service. The primary aim here is not of itself to eliminate lawyers, although this may to some extent be the consequence. The primary aim, the driving imperative, is to meet the 'more-for-less' challenge as outlined in Chapter 1. The primary aim is to help clients.

I say again: most mainstream lawyers and judges cannot imagine a revolution being initiated by any start-up (or indeed upstart). At root, this is a failure of imagination.

My second category of game-changing lawtech start-ups will help solve the access to justice problems (see Chapter 12).

There is some overlap here with the first category, in that some traditional providers do indeed strive to offer and increase access. But the gargantuan opportunity here is to develop systems and services where today legal and court service is beyond the reach of humanity. The challenge here is to ask how can we use the remarkable and steadily increasing reach of the internet, mobile technologies, and artificial intelligence to help many more people to understand their legal rights and enforce these with the backing of the law and courts? We will need technologies that can provide online legal guidance of various sorts as well as technologies that can enable the conduct of judicial work remotely on an online basis—by video hearing and by asynchronous processes (see Chapter 14). I have confidence that some start-ups will meet these needs and their systems and services will in turn liberate the 'latent legal market' (Chapter 15).

I do put much of my faith in start-ups to deliver these game-changing solutions. In terms of ethos, I find the lawtech start-up community to be the most energetic and inspiring corner of the legal world. Free from the shackles of tradition, the founders should be able to imagine legal services that are quite unlike today's. They can challenge conventional assumptions. They can disrespect current boundaries while maintaining respect for the law. Above all, they can be driven by the needs of those who need legal help and support rather than by those who are the incumbent providers.

I do not expect there will be a single Uber or eBay in law (the legal field has too many branches and service types to be reduced to one system or app), but I do expect that a variety of breakthroughs will in combination transform the legal market.

11 | The Timing of the Changes

I am often asked to indicate the likely time scales of the changes that I predict. Some commentators and lawyers believe these changes are already upon us and that the legal world will have transformed within a couple of years. Others maintain that the shifts will progress at a more plodding rate and that it will be some decades before the revolution is complete. I believe that we will witness neither a big bang revolution nor a slow burn evolution. Instead I have always expected an incremental transformation to unfold, in three stages—denial, re-sourcing, and disruption (see Figure 11.1).

I am not suggesting that all law firms and in-house legal departments will move evenly through these three stages in concert. Some pioneers will progress far more rapidly, while there will be laggards aplenty who will take much longer to advance. Like all models, this one is a simplification. Its purpose is to give a broad sense of the order in which most large law firms and in-house legal departments will move forward. The speed with which they move cannot be predicted with precision; this will depend largely on factors such as the state of the economy, the intensity of the demands made by clients, the impact of

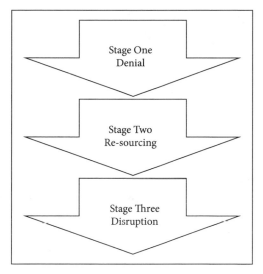

FIGURE 11.1 *The three stages of change.*

new competitors on the market, and whether or not a few firms take a positive lead in changing the way that legal services are delivered.

Stage 1—Denial

This stage ran from around 2007 to 2016. During this period, most lawyers, both in firms and in-house, wished it were 2006 again. By 2016, however, very few legal leaders with whom I met were denying that fundamental change was afoot. It was widely acknowledged that the old times had passed—the era when many law firms had more work than they could handle, with the added benefit that it was not price sensitive.

As for in-house legal departments, although there were hints before the global recession (2007–2009) that they would need to think about tightening their belts, they were generally not under great pressure within their organizations to spend less. Before the financial crisis, it was a time of abundance. It was a sellers' market and the buyers' purses were brimming (or so it seemed).

In 2007, with the economic downturn and then crisis, came the clear start of the more-for-less challenge, as introduced in Chapter 1, and we were also at the beginning of almost a decade of denial. The initial response of many General Counsel (GCs) was to seek to reduce their legal expenses, not by fundamentally changing their own internal operations and sourcing strategies but by asking their external law firms for significant reductions in fees. This was done fairly formally in many cases through competitive tendering processes. In turn, most law firms proposed a variety of alternative fee arrangements (AFAs) while also making many lawyers (from junior lawyers through to partners) and support staff redundant, and spending less on back-office functions such as technology and marketing.

In truth, most in-house legal departments and law firms were eager during this stage of denial to weather the storm without major upheaval: in-house lawyers hoped to maintain their headcount, while law firms strove, through (non-fundamental) cuts, to maintain their profitability.

During this first stage, some law firms engaged in tokenism of a kind—they partook of some modest alternative sourcing, but largely as a matter of show. Their purpose was to construct a credible narrative to relay to those clients who inquired as to their cost-cutting strategies. The deeper purpose was to play for time. This was a stalling strategy, adopted in the hope that the market would return and clients would no longer have an urgent need to spend less on law firms.

At the same time, some in-house lawyers argued that the answer for them was to increase the size of their in-house departments on the ground that in-house lawyers cost less than legal specialists in law firms. This often proved to be tactically flawed. It is an approach that can make unrealistic assumptions about how fully occupied the in-house lawyers will be. Also, supply often increases demand so that recruiting one in-house lawyer often leads to that person asking for an assistant and then a team. And, in any event, growing the in-house capability is often to fall back on the outmoded view that there are only two ways of sourcing legal work—within a department or through a law firm—and so ignores the possibility of alternative sourcing.

In summary, in this first stage, most (but by no means all) lawyers denied that fundamental and structural changes were at play within the legal marketplace. They assumed, or prayed, that when the economic tide turned and business improved, the legal world would return to its modus operandi of around 2006. This did not happen, largely because those who were running client businesses saw that legal costs could be managed more tightly, that legal work could be undertaken differently and more efficiently, and so they had no appetite for a return to the old inefficiencies.

Stage 2—Re-sourcing

During the stage of denial, chief executives and boards noticed that legal costs did not always plummet as requested. Although GCs had spoken optimistically to chief finance officers about reductions in hourly rates and alternative fee arrangements, it became clear that the promised reductions were not materializing. Lawyers, both in-house and in firms, recognized that they needed to move from pricing differently to working differently. This was a move from the first to the second stage. Most legal organizations are still in this second stage, while some are edging towards the third.

It has not sufficed, in this second stage, for GCs to ignore the inefficiencies in their own departments. Just as they have asked law firms to find alternative ways of sourcing the routine work that used to be done at high hourly rates by junior lawyers, so too they have needed to apply this same approach within their own legal businesses. Both law firms and in-house legal departments have therefore analysed their legal work and identified ways in which the most straightforward, procedural, and administrative-based activities and tasks could be sourced differently, whether by outsourcing, off-shoring, using paralegals, computerizing, or deploying any of the various sourcing strategies laid out in Chapter 5. In-house departments have also begun to show an interest in collaborating more with one another, in sharing the costs of legal services in the manner anticipated in the collaboration strategy (Chapter 3).

Also during this stage, the alternative providers of services to the legal profession—legal process outsourcers, publishers, law companies, accounting firms, private equity-backed start-ups, and many more—have come to play a more significant role in the delivery of legal services. There has never been a tipping point, but we have begun to witness an upsurge in new competition for law firms.

Additionally, in this second stage, many law firms and in-house departments have found ways of running their own back offices (technology and accounts, for example) at far lower cost. Some have resorted to business process outsourcing or shared services facilities.

In this second stage, there has also been a growing interest by in-house legal departments in legal needs analysis and legal risk management. Owners and leaders of businesses have wanted to have a far clearer sense of their urgent and unavoidable legal expenses and to move towards confining their legal spend to these alone.

Stage 3—Disruption

In Stage 2, alternative sourcing has generally been achieved by requiring legal work to be undertaken by lower-cost labour. Effectively, this has been a form of labour arbitrage, and it has resulted in notable savings. However, having less costly people undertake legal work, with playbooks or basic automation, is certainly not the end-game in the evolution of the legal marketplace.

Some more advanced firms and departments have moved into Stage 3, which will bring much more radical transformation. This will be the path followed by the majority of successful legal businesses of the 2020s, and will be brought about by the introduction of increasingly capable systems. In the main, these technologies will be disruptive, that is, they will challenge and displace the traditional way in which legal work has been done in the past.

Many lawyers and commentators fail to recognize that services such as legal process outsourcing (LPO) and managed legal services, where they largely use low-cost labour, were always destined to be temporary measures and not long-term solutions. If we think of two classes of service that are typically considered to be suitable for outsourcing beyond law firms and legal departments—document review in litigation and basic contract drafting—both of these will in due course be replaced by capable systems that can outperform junior lawyers and administrative staff. The very features of these areas of legal work that make them suitable for outsourcing—that they can be broken down into manageable parts, and that they can be discharged by well-trained but relatively unskilled human beings with the assistance of detailed procedures or playbooks—are also precisely the features of work that make them amenable to the application of technology. Thus, technology-assisted review used for electronic disclosure can comfortably outperform many human beings who conduct document review in litigation. (Electronic disclosure, or e-disclosure, broadly speaking, involves parties to a dispute letting one another know about the existence of relevant, electronically stored documents.) And, likewise, document automation already operates more reliably and efficiently than modestly experienced lawyers and administrators.

The widespread and pervasive deployment of disruptive technologies represents the end-game for legal service although, even then, as noted in Chapter 1, there is no finishing line in the world of technology. In the long run, increasing amounts of legal work can and will be taken on by highly capable systems, with light supervision by the human beings who are their users.

This will be the context and backdrop of the careers and working lives of tomorrow's lawyers.

These disruptive technologies will come to dominate not just in substantive legal work but also in the way that providers of legal services (both humans and systems) are selected. Price comparison systems, reputation systems, and online auctions for legal services (see Chapter 6) will be used frequently, creating an electronic legal marketplace quite unlike the traditional basis of legal trading that has endured for decades and more.

It is not that capable systems will replace all legal work by, say, the mid-2020s. Of course not. But around that time and from then on it will become commonplace across the legal profession for all substantial and successful legal businesses to be converting their business processes from human handcrafting to ever more sophisticated and capable technology-based production. We have seen such changes in many other sectors of our economy and there is no reason to think that the law should be immune from technology. If analogous technologies can transform the practice of medicine and audit, then lawyers should be open to similar overhaul.

A legal world will emerge that is manifestly different from today's. And it is into this world that most young lawyers will be stepping. For those aspiring lawyers who hoped for a career akin to that enjoyed by lawyers of their parents' generation, they will be disappointed. For those who seek new opportunities and wish to participate in bringing about the advances that I predict in this book, I believe there has never been a more exciting time.

12 | Access to Justice and Online Legal Services

In 2000, in my book *Transforming the Law*, I predicted that within five years, more people in the UK would have access to the Internet than access to justice. Unfortunately, I was proven correct. Today, as I explain shortly, less than 3 per cent of British people are effectively excluded from the Internet, whereas the majority of citizens in England and Wales are unable to afford most of the services of lawyers and of the courts. The good news, as I have for long claimed, is that technology will be pivotal in overcoming many of the growing problems of access to justice. My aim in this chapter is to show how this might come to pass.

Access to Justice

Franz Kafka sets the scene hauntingly in *The Trial*. He tells of a gatekeeper who inexplicably refuses to grant a man access to the law. This unfortunate man from the country had not expected any problems. After all, he thinks, 'the law should be accessible to everyone at all times'. So it might be thought, but research in England and Wales conducted a few years ago concluded that around 1 million civil justice problems go unresolved each year. This legal exclusion or unmet legal need is a grave social problem and is loosely referred to as the 'access to justice' problem.

Thinking more widely for a moment, no-one today can pretend to have mastery over anything other than small parts of our legal system. And yet every one of us, under the law, is taken to have knowledge of all legal provisions that affect us. Given that most citizens do not know most of the law and cannot afford to obtain conventional legal advice, we seem to be in a rather parlous state. The problem perhaps comes most sharply into focus when people contemplate taking an action through the court system. From a lay perspective, as well as appearing to be unaffordable, the courts seem to be excessively time-consuming, unjustifiably combative, and inexplicably steeped in opaque procedure and language. It was with such problems in mind that in 1995 and 1996, Lord Woolf, then a Law Lord and later the Lord Chief Justice of England and Wales, published 'Access to Justice', his two seminal reports on the future of the civil justice system. Lord Woolf's terms of reference confined his attention to the *resolution* of disputes. And for many judges and policy-makers since, the idea of improving access to justice has come to mean improving and accelerating the way disputes are resolved in courts.

I take a wider view. To be entirely or even substantially focused on dispute resolution in our pursuit of justice is, I claim, to miss much that we should expect of our legal systems. It is my contention that better access to justice should embrace improvements not just to dispute resolution but also to what I call dispute containment, dispute avoidance, and legal health promotion.

Dispute containment concentrates on preventing disagreements that have arisen from escalating excessively, and it is lawyers as well as the parties themselves who need to be contained. Dispute avoidance is a theme that in-house lawyers often raise with me: they speak of legal risk management or, as I put it, putting a fence at the top of a cliff rather than an ambulance at the bottom. I have yet to meet a regular human being, whether a chief executive or a consumer, who would prefer a large dispute well resolved by lawyers to not having one in the first place. Legal health promotion extends beyond the preventative lawyering of dispute avoidance to ensuring that people are aware of and able to take advantage of the many benefits, improvements, and advantages that the law can confer, even if no problem has arisen.

Recognition, Selection, and Service

With these various dimensions of access to justice in mind, the plot thickens somewhat when we reflect on the plight of the non-lawyer. (Parenthetically, a word about the term, 'non-lawyer' because people get very exercised about the label. I wholly agree that it is generally disparaging and unacceptable within a law firm to pigeon-hole an entire community of employees in terms that suggest they are second-class contributors. Categorizing humans primarily in terms of what they are not is bad practice. However, to speak of non-lawyers in a much wider context, beyond the confines legal organizations, can be a powerful and not disrespectful way of highlighting that the group to which it refers (most of humanity) are not knowledgeable about the law.)

The first obstacle for the non-lawyer (in this latter sense) is *recognition*, the process by which someone with no legal insight realizes that he or she would benefit from legal help. Sometimes it is obvious—when a claim arrives through the letterbox or a decision has been made to move house. But often non-lawyers may not know that they are in a situation in which there is a legal problem to be resolved, contained, avoided, or that there is some benefit to be secured. Paradoxically, it seems you need to be a lawyer to know if and when you would benefit from legal help.

The second challenge, even if our non-lawyer has recognized that he or she would benefit from legal help, is to select the best source of legal guidance, whether that be finding a suitable lawyer or some other kind of adviser or even online help.

The third dimension is the delivery of legal service itself. And here we find the wide range of options that people now face in what I call a multi-sourcing environment (see Chapter 5). In relation to this third challenge, I do not believe that conventional lawyers in traditional law firms are always the best placed or most affordable sources of guidance for clients. It seems almost inevitable that ongoing cuts in public legal funding, brought about in part by economic conditions, will lead to legal and court services that are steadily less affordable and less accessible. A major and urgent social challenge is to find new ways of providing legal help, not least to citizens and to small businesses.

Online Legal Services

One clear alternative to the provision of legal help by lawyers is for trained and often voluntary non-lawyers to advise people on their problems, rights, and responsibilities. In the UK, for instance, the Citizens Advice service does precisely this but it too suffers from lack of resources. Another option is to provide citizens and businesses with online legal resources so that they can take care of some of their legal affairs on their own; or, when guidance is needed, they can work more efficiently with their legal advisers. If we can have state-provided online services that provide broad medical guidance (e.g., at <http://www.nhs.uk>) why not have something similar for law?

Sceptics should bear in mind that online services generally are no longer just for a high-tech minority. On the contrary, the Internet is now central to the lives of the majority of families and businesses in England and Wales. Most studies suggest that around 95 per cent of the British population are now Internet users of one sort or another. The remaining 5 per cent are, of course, important, but as research at the Oxford Internet Institute (its 'Internet in Britain' studies) has pointed out, only a fraction of these non-users (or ex-users) 'definitely don't know' someone who could assist them. In practice, this means less than 3 per cent are currently out of reach—fewer than one in 30 people are non-users who have no-one who can use online services on their behalf. This is a smaller percentage than is often supposed.

As for online legal facilities, these currently come in three main forms: first, as free Web-based services, provided by a variety of commercial and not-for-profit organizations; second, as subscription-based tools from conventional law firms; and, third, as chargeable offerings from other businesses, such as alternative business structures or legal publishers.

In practice, then, how might online services actually help to secure access to justice in all the various aspects I note? In the first instance, addressing the initial obstacle noted earlier, technology can and will continue to be of use in assisting non-lawyers to recognize that they might benefit from some kind of legal input. One approach will be for people to register their social and working interests and for legal alerts to be delivered automatically to them when there are new laws or changes in old law that apply to them. Another tack will be online triage—when a citizen has a grievance of some sort, a simple online diagnostic system could ask a series of questions, require some boxes to be ticked, and could then identify if the user has a legal issue, and if so, of what sort.

A further possibility, as mentioned in Chapter 6, will be the embedding of legal rules into systems and procedures. Consider the game of Solitaire. When I was a boy, we played this with atoms (playing cards). It was possible, when using these cards, to place a red 4 beneath a red 5, although this would clearly have been in breach of the rules. But it would have been physically possible. In contrast, when you play Solitaire on your computer, such a move is not possible. Any attempt to place a red 4 below a red 5 will be met with a refusal by the system to do as clicked. The difference here is that, with the electronic version, the rules are embedded in the system. Failure to comply is not an option. In years to come, in many dimensions of our social and working lives, I predict that legal rules will similarly be embedded in our systems and processes. This means that non-lawyers will no longer have to worry about, or have the responsibility of, recognizing when legal input is required.

A final use of technology to help non-lawyers recognize when they need legal help will be through what I call 'communities of legal experience'. As a computer user, when confronted with some incomprehensible error message, you will no doubt have cut and pasted the message into Google, and found that someone out there has already provided an explanation and solution to your problem. So too in law, I believe that in open source and wiki spirit, large communities of legal experience will build up so that people will learn of legal issues that affect them, not formally through notification by their lawyers but informally through their social networks.

Technology will also play a role in helping clients to select their lawyers and other sources of guidance. As explained in Chapter 6, there will be online reputation systems, not unlike those services that offer collective feedback on hotels and restaurants, which will provide insight from other clients into their experiences with particular law firms and lawyers. There will also be price comparison systems that will allow non-lawyers to assess the respective prices of competing legal providers. And there will be auctions for legal services—not generally for complex bespoke work but for the routine and repetitive work that I say will be sourced in various ways in the future.

As for the role of technology in the actual delivery of legal service, increasingly people will turn for basic guidance, on procedural and substantive issues, not to lawyers but to online legal services. We already use so much online information in our daily lives that there is no reason, especially for those who cannot afford otherwise, why legal help should not be similarly accessed. Equally, users will turn to online services for the production of standard documents, such as basic wills and landlord and tenant agreements (see Chapter 4), to diagnostic systems that can pinpoint their rights and duties, and to communities of legal experience to determine how fellow lay people have sorted out their difficulties in the past.

Another set of possibilities, as described in Chapter 14, are online courts and online dispute resolution (ODR). Whether through public court services delivered online or via private-sector forms of ADR (alternative dispute resolution), such as e-mediation and e-negotiation, non-lawyers will themselves be able to understand and enforce their entitlements, and so resolve their differences without the costly involvement of lawyers.

Yet another possibility will be to build social networks of lawyers or legal advisers who are willing, in their own time rather than on a face-to-face basis, to provide online guidance, in a variety of ways (either directly to citizens or indirectly to advice workers).

Although I speak of many of these systems as belonging to the future, there are already innumerable examples of operational online legal services. In the words of William Gibson, the science fiction writer, 'The future has already arrived. It's just not evenly distributed yet.' It is early days to be sure, but within a small number of years these systems will be commonplace in helping non-lawyers to recognize when they need legal help and to select the best sources of advice, as well as in actually offering them practical guidance. And this is not just the pipe dream of some Internet enthusiasts. Significantly, empirical research is frequently finding that there is considerable enthusiasm amongst consumers for the online delivery of reliable legal support and advice.

Some of these uses of online legal services will be 'disruptive' for traditional law firms, in the sense discussed in Chapter 6. But, at the same time, many of these techniques will make the law available to people who would otherwise have no affordable sources of legal help. This I call the realization of the 'latent legal market'—those countless occasions in the lives of many people when they need legal help and would benefit from legal help but, until now, they have been unable to secure this assistance (whether to resolve, contain, or avoid problems, or indeed to afford them some benefit). As Chapter 15 and the arguments just made suggest, online legal service is liberating the latent legal market.

13 | Judges, Courts, and Technology

In 1981, while studying law at the University of Glasgow, I wrote an undergraduate dissertation on computers and the judicial process. My interest was in the extent to which the work of judges could be supported or even replaced by advanced computer systems. The potential and limitations of judicial and court technologies continue to fascinate me and I have had the good fortune over the years to be able to explore my thinking further in collaboration with a number of England's most senior judges, especially in my capacity as Technology Adviser to the Lord Chief Justice, a position I have held since 1998.

Judges and Technology

Judges are commonly portrayed, by the media and in fiction, as old-fashioned and otherworldly. It might therefore be expected that judiciaries in advanced jurisdictions are made up of the last of the neo-Luddites. The reverse is so. Most judges with whom I work and speak are now committed users of technology and are keen to embrace all basic systems that offer practical benefits in their everyday work, such as email, word processing, and online research.

Looking beyond these rudimentary applications, how profoundly could technology affect the work of judges? In this chapter, I focus on what is available and practicable today, while in the next I explore some cutting-edge systems.

Before delving into technology, I should stress that some of the more general techniques and lessons of this book can be applied to judges as much as to other lawyers. Consistent with the ideas of decomposing and multi-sourcing as introduced in Chapter 5, I see great scope, for example, for analysing and dividing judicial work into separate parts and, where appropriate, finding alternative and more efficient ways of undertaking some of these tasks. Judges frequently tell me that they are called upon to undertake mountains of administration that others, less qualified, could handle on their behalf. At the same time, there is scope for standardization of at least parts of the documents (the directions and orders, for instance) that judges create. More, the judiciary would clearly benefit from the use of document automation, where much of what appears in these final documents is standard wording with minor variations (see Chapter 4).

Initial legal research could also be conducted in different ways, as has already been demonstrated to some extent in England and Wales by the deployment of judicial assistants in the Court of Appeal and in the Supreme Court. While it would not be feasible to use these junior lawyers across the entire system, there are other innovative ways, using technology, in which know-how and experience could be shared to similar effect. I am calling, therefore, at the very least, for serious, further investigation of the scope for the decomposing and multi-sourcing of judicial work. So far, senior judges have met this call open-mindedly and so I expect change in the direction noted.

Disappointing Progress

As for technology, most current and emerging systems for judges are 'sustaining', in the language of Chapter 6. They support and enhance the way that judicial work is currently undertaken. In the next chapter, I look at technologies that are more 'disruptive'—challenging and even replacing today's practices and processes.

One area in which there is judicial disappointment in most jurisdictions is in applications that fall under the broad heading of 'case management'. This term is used in many different ways, but the generic grievance here is that the systems that are available to judges do not generally support them directly in the everyday management of their cases and documents.

Many judges, in the name of case management, hanker after a 'digital case file', the idea of which is that all documents relating to particular cases are submitted (e-filed) to the courts in electronic form and are available to the judges and officials as electronic or digital bundles. This is not rocket science. It is barely computer science. But it should be commonplace and deeply entrenched in a twenty-first-century justice system.

Other judges rightly expect more—as well as digital bundles and files for better management of the documents, they want technology-based workflow or project management systems to streamline and enhance standard processes (see Chapter 6) and so drive cases through the system more rigorously, consistently, and efficiently.

In most countries, despite the remarkable advances in technology generally, progress in the past 20 years in digital case management has been slow. The organization of most of the work of the world's courts is still labour-intensive, cumbersome, and paper-based. Although remedial action is under way (as discussed shortly), a visit to most courts in England and Wales still reveals a working environment that is less efficient and automated than most ordinary offices in the country, whether in the public or private sector. Across the country, judges complain of antiquated systems, outdated working practices, excessive running costs, inefficiencies, errors, and delays. In turn, court users suffer and the reputation of the justice system is adversely affected.

England and Wales

In his 'Access to Justice' reports, Lord Woolf made a series of recommendations in the mid-1990s for the computerization of much of the case management in the civil justice system of England and Wales. To this day, few have been implemented. Until recently, this lack of progress was attributable to two main factors: insufficient investment by the government and the Treasury, which did not consider civil justice to be a priority; and the Ministry of Justice's poor track record of successfully procuring and delivering large-scale technology projects. There was no shortage of vision from within the justice system. Over the years, a number of enlightened judges, politicians, and civil servants have expressed bold views of a court and justice system transformed through technology. In short, and this is a global story, there was sufficient vision but insufficient cash and technological capability.

However, a new era began to dawn in England and Wales in 2014, when there was a major drive—led by senior judges, politicians, and officials—to embrace technology across the court system. Much work was undertaken behind the scenes during 2014 and 2015, so that it came as a surprise to many practising lawyers when, on 25 November 2015, the British government announced it was to invest 'more than £700 million to modernize and fully digitize the courts'. My reaction on Twitter ran as follows: 'I have waited for this day for 34 years'. At last, after innumerable under-funded false starts, there seemed to be serious commitment to upgrading the court system.

The modernization and reform programme, it was claimed, was not just about technology—there were also plans to reduce the number of court buildings and to transfer some work of judges to legal officers. But technology was always at the heart of the initiative, including systems to create a common platform for the prosecution service and courts (the fourth attempt to do something similar in the past 25 years); case management systems for criminal, civil, and family work; e-filing across the entire court and tribunal system (see later in this chapter); virtual hearings (ditto), online courts (see Chapter 14), and much else besides. Old systems were to be scrapped. Tomorrow's courts were to be built on the back of technology.

This ambitious programme received formal approval in a joint vision statement by the government and the judiciary, entitled 'Transforming Our Justice System' (September 2016). This paper declared a shared commitment to 'a courts and tribunal system that is just, and proportionate and accessible to everyone'. In the event, the programme has cost north of £1 billion, it has not yet been completed, and interim formal reviews have in varying degrees been critical. It is due to finish in 2023 but it is not clear at the time of writing (mid-2022) what the final and full configuration of systems will look like, nor indeed whether the systems will meet users' expectations and needs.

The Way Ahead

Whatever the outcome of the reform programme in England and Wales, the arguments in favour of change—and change well beyond the current reforms—remain compelling and are of general applicability across all jurisdictions. Our court system continues to creak. Too often, it is inefficient, slow, too costly, and beyond the ken of the non-lawyer. As noted in Chapter 12, around 1 million civil justice problems in England and Wales are said to go unresolved every year and cuts in legal aid have added greatly to this shocking level of legal exclusion. 'Access to justice' is in grave danger of being a good that is available only to the rich. More than this, full-scale civil litigation at disproportionate expense is, too often, wielding a sledgehammer to miss a nut.

In principle, if the advantages of technology that are seen in other sectors are enjoyed by the courts, the labour-intensive, cumbersome, and paper-based systems for court administration will be replaced by an automated, streamlined, and largely paper-free set of systems that will be less costly, less prone to error, more efficient, and more accessible. This would be a system that is easier to use than today's and in large part delivered online. The processes of starting a case, communicating with the court, storing documents, monitoring progress, arranging hearings, and receiving support along the way would be conducted electronically.

In turn, an efficient, intelligible, and well-equipped court service, populated by satisfied lawyers, would be a system in which the public would have greater confidence. There is an international dimension too. If England seriously aspires to being a leading global centre of excellence for the resolution of disputes, then there should be state-of-the-art, leading-edge systems, processes, and infrastructure in place to support this.

That is why our government must continue to fund technology in support of the courts and dispute resolution. The current preoccupation of the state, understandably, may be with cutting costs rather than investing in technology. But the paradox here, in this period of economic gloom, is that investment in technology should be a solution for governments and not a problem. The prize is a glittering one—inexpensive, swift, proportionate, inclusive resolution of disputes.

Technology-Enabled Courts

Looking beyond back-office administration and case management, what is now technologically possible in the courts themselves? (I apologize to any technologists from beyond the law who are horrified that some of the technologies in this chapter are even thought worthy of mention in the third decade of the twenty-first century.)

First of all, even before parties assemble in court, there is one technique that can be of immediate benefit—e-filing. This involves the submission of documents to the court in electronic form, which can be so much more convenient for judges and administrators than huge bundles of paper, especially if the e-filed pages are linked to one another. Over the past few years, in the UK and elsewhere, e-filing has become increasingly popular.

Next, in the courtroom itself, one entirely obvious but by no means pervasive use of technology involves the judge taking notes as cases progress on a laptop or desktop computer. Three further, more sophisticated, technologies have been used, to a greater or lesser extent, in courts around the world since the early 1990s. The first is computer-assisted transcription (CAT), which enables words spoken in the courtroom to be captured by stenographers and then converted into text that appears almost instantaneously on the screens of judges and other participants. The text can be annotated as it appears and a searchable database of the proceedings is created. (It is only a matter of time before CAT is replaced by continuous speech, speaker-independent, multi-speaker voice recognition systems.)

Second are document display systems, which ensure that everyone in a hearing is, literally, on the same page—instead of waiting for all parties and judges to locate papers and files manually, the court's attention can be directed to a particular page by asking all participants to look at their monitors, which instantly display relevant documents.

Research and experience suggest that using CAT and document display technologies reduces the length of hearings by one-quarter to one-third.

The third technology is digital or electronic presentation of evidence and reflects the old adage that a picture is worth a thousand words. Rather than relying exclusively on oral advocacy, lawyers can present evidence using a wide range of non-verbal tools, including charts, graphs, diagrams, drawings, models, 3D animations, reconstructions, and simulations. These can be displayed in the courtroom on individual monitors or projected onto very large screens. They can be used in both civil and criminal cases—for instance, the extent of a delay in a project can be demonstrated powerfully by an animation that compares actual with projected time taken, or complex movements of funds can be captured in a simple graphic rather than by convoluted oral summary.

As more and more evidence itself is created or preserved in digital form, a further use of technology is devoted to managing this electronic evidence and integrating it with court practice and procedure. This includes emails, CCTV footage, and data stored on a blockchain. This last category is increasingly important because evidence stored and generated from a blockchain carries with it powerful claims to authenticity, integrity, and accuracy.

As for actual use of CAT, document display systems, and digital presentation of evidence, here too take-up has been low across court systems, despite the time and cost savings. But there have been a few notable exceptions. Even before the pandemic, the UK Supreme Court was on to its second generation of systems for e-filing, document display, real-time transcription, judicial use of computers on the bench, and remote evidence.

And various tribunals have impressive systems. These, however, are exceptional in the UK, where most hearing rooms, aside from laptops, are substantially technology-free.

Interestingly, court technologies have enjoyed greatest success in public inquiries (where the resources for setting up the hearing rooms have been less constrained—still the most ambitious in this category was the high-tech Bloody Sunday Inquiry (1998–2010) by Lord Saville), in large-scale commercial disputes (where the parties themselves have substantial litigation budgets); and complex criminal cases (the criminal justice system has, historically, received much greater investment in its technology than the civil justice system).

In the future, many courtrooms will look like Lord Saville's Inquiry hearing rooms; not unlike NASA control centres.

Remote Courts

Turning now to the heart of dispute resolution, in thinking about the long-term future of courts, one fundamental question sets the agenda: is court a service or a place? To resolve legal disputes, do parties and their advisers need to congregate together in one physical space, in order to present arguments to a judge? What about more advanced forms of online dispute resolution?

In the second edition of this book, published in 2017, I wrote that, '(v)irtual hearings are still relatively rare across the justice system in England, but they will become commonplace in the 2020s, I have little doubt.' In the event, the pandemic brought about changes that confirmed this prediction much sooner than I had expected. Prior to this, there had been minimal use of video-hearing technology. Take-up had been greatest in criminal cases, with child or intimidated witnesses; and for bail and remand hearings, conducted through links between prisons and courts. And in some civil cases, witnesses from outside the UK had given evidence remotely, as had otherwise inaccessible expert witnesses. In other words, video hearings had been in the peripheral vision of judges, lawyers, and policy-makers until 2020.

And then came COVID. Courtrooms around the world closed. Faced with no feasible alternatives, judges, lawyers, and court officials moved with extraordinary speed and agility to conducting hearings remotely. In March 2020, I led a team that put together a website known as Remote Courts Worldwide—<http://www.remotecourts.org>—to help the global community of justice workers—judges, lawyers, court officials, litigants, court technologists—to share their experiences of remote hearings. In mid-2022, Remote Courts Worldwide records that 170 jurisdictions have run remote hearings in the past couple of years. This scale of uptake was scarcely imaginable at the start of 2020.

The term 'remote' generally refers to video hearings (using systems such as Zoom or Teams, or custom-made services) or to audio hearings (essentially, telephone conference calls). There are hybrid versions too but in practice, the video hearing quickly emerged as the favoured option during the pandemic. Instead of assembling in a physical space, participants convene in a video meeting. The judges, lawyers, and the procedures remain much the same.

What lessons have we learned so far from the widespread use of remote courts? It should be accepted immediately just how flexible and resourceful lawyers and judges can be when circumstances require. The old trope is that the legal profession is deeply conservative and incapable of change. The evidence that appears on Remote Courts Worldwide suggests differently.

Overall, the level of satisfaction with video hearings amongst judicial and legal users is fairly high and much higher than they would have predicted if asked before the pandemic about the potential of remote hearings. There have of course been wrinkles and problems. There have been difficulties, for example, for elderly participants, for those needing translation, and for court users with limited technology. There have also been concerns about privacy and security on some of the video platforms. It turns out to be difficult to handle large numbers of documents in a video hearing if the court system itself is still largely paper-based. If solemnity is to be maintained, then lighting, background, clothing, tone, and posture matter. Low-quality Internet connection can scupper proceedings entirely.

There remain some issues of debate, in relation to the credibility of witnesses, for instance, and whether this can be adequately assessed in video hearings. Some say you need to be in the same room as humans to determine whether they are speaking the truth. Others insist that you can still look people in the eye when engaging remotely.

There is agreement that more work needs to be done to establish what kinds of cases or issues are best suited to what types of disposal, whether by physical, audio, video, or by other online methods (see Chapter 14). Some early positions, though, are gaining traction. It is widely agreed, for example, that family disputes, involving the custody of children or domestic abuse, should be heard in person if possible. Likewise, in most countries, cases involving serious crime are thought to require physical hearings. On the other hand, particular sorts of dispute are emerging as nicely suited to remote handling—interim, procedural, and interlocutory hearings; routine family work; small money claims; minor criminal offences; commercial disputes; administrative tribunals; civil appeals; and more.

In sum, the research and feedback we have seen so far indicates that some—and probably many—legal disputes can indeed be handled remotely, often more cheaply, conveniently, speedily, and affably than in our traditional system. But this can only be taken as a tentative hypothesis that must be challenged and tested systematically.

Many lawyers and judges are nonetheless insisting we should never return to the old ways, that the shift to technology-enabled justice has been achieved. Job done, allegedly. This is overclaim. Incontestably, the jump from physical courts to remote hearings has been extraordinary, but it is still early days and no-one can responsibly assert that video is suitable for all cases and issues. We are, I insist, at the foothills of the transformation in court services. The current assortment of remote services comprise a brave array of *ad hoc* services but much effort and capital will be needed to industrialize these first-generation systems to construct court services that are scalable, stable, and, vitally, designed for use as much by lay people as by lawyers.

More than this, the systems that have been cobbled together are still illustrations of what I call, in Chapter 1, *automation* rather than *innovation*. Most of the remote services that were set up in response to COVID are variations on the theme of conventional courts. But we should be clear—dropping our current court systems into Zoom is not, as some pundits like to proclaim, a 'revolution'. We have just scratched the surface. The future of dispute resolution lies beyond technology-enabled physical courts and video hearings. The future will be a world of online courts and online dispute resolution, to which I now turn.

14 | Online Courts and Online Dispute Resolution

In a video hearing, one or more judges dispense justice in the traditional manner. The break from tradition is that some or all participants appear virtually across some video link rather than in person. This, as noted in the previous chapter, was the most popular form of remote hearing during the pandemic.

Advances are now being made well beyond virtual hearings into the world of online courts and online dispute resolution (ODR). Some experts are going further and giving thought to the idea of legal hearings in the metaverse. In these brave new worlds, no traditional courtroom is involved. Instead, the process of resolving a dispute, especially the formulation of the solution, is entirely or largely conducted online. A court becomes a service rather than a place.

Online Courts

The story of online courts is unfolding as I write. They may well have greater impact on everyday life than any other innovation discussed in this book. Much that is said in this chapter is laid out at greater length in my book, *Online Courts and the Future of Justice*, which first appeared in 2019, but required an update in light of COVID. A revised edition appeared in 2021.

In the book, I propose a fairly detailed architecture for online courts. Distilled to its basics, there are two aspects to online courts, as I see them. First of all, there is *online judging*. This involves the determination of cases by human judges but—and here is a fundamental departure—the parties neither assemble in a bricks-and-mortar court building nor are proceedings conducted by video or telephone conference call. Instead, arguments and evidence are submitted to judges through some kind of online platform and the judges deliver their decisions, again, not in a video hearing or in open court but via the online service. There is no hearing or oral argument. Instead, there is a structured process of online pleading, not unlike an ongoing exchange of emails and attachments.

In the jargon of technologists, interaction in physical courtrooms and video hearings is *synchronous* whereas online judging involves *asynchronous* forms of communication. For the former, this means that the participants need to be available at the same time for a case to progress. By contrast, with the latter, as with exchanging text messages and emails, those who are involved do not need to be at hand simultaneously—evidence, arguments, and decisions can be transmitted and traded without the sender and recipient being physically together or connected by video at the same moment.

This move from a synchronous to an asynchronous court service is not mere process improvement. It involves and requires radical change. It constitutes a much greater shift than the swing from physical to video hearings. Although critics have deep misgivings about video hearings, they belong, as a form of dispute resolution, in the same broad paradigm as traditional courtrooms. By contrast, online courts are an entirely different concept. Even in the first generation of online courts, where human judges are determining the cases, online judging removes much that many people hold dear—the public hearing, the day in court, the direct engagement with other human beings (more on which shortly). On the other hand, as I and many others strongly believe, online courts are likely to make state-based dispute resolution much more accessible and affordable, and will chime with the digital generation, those who cannot recall a pre-Internet world.

The second aspect of online courts is more general. I refer to it as *the extended court*. Here, the idea is that technology enables us to provide a service with much wider remit than the conventional court. The additional services include tools to help users with grievances to classify and categorize their problems, to understand their rights and obligations, and to understand the options and remedies available to them, as well as facilities that assist litigants to marshal their evidence and formulate their arguments, and systems that advise on or bring about non-judicial settlement. Everyday techniques and technologies—apps, smart phones, portals, messaging, video calling, chat bots, live chats, webcasts—will support these extended court facilities, helping non-lawyers interact much more easily with the courts. The extension here, and it is a major change, is that these systems are designed primarily for self-represented litigants rather than for lawyers. And these court users, without lawyers, can themselves file documents, track cases, engage with court officials and judges, and progress their disputes by using intuitive, jargon-free systems.

Impact

Two major benefits can flow from online courts as I describe them—an increase in access to justice (a more affordable and user-friendly service) and substantial savings in costs, both for individual litigants and for the court system. In the language of Chapter 12, for appropriate classes of case, online courts not only promise more accessible and affordable dispute resolution but, crucially, their extended court facilities also encourage dispute containment and dispute avoidance.

Online courts are not science fiction. They are operational in Canada (the Civil Resolution Tribunal in British Columbia is the best global case study), in the United States, in China, and elsewhere.

The introduction of online courts in England and Wales is also part of the government reform programme mentioned in Chapter 13. I think it fair to say that the drive in this direction originated largely from the work of the Civil Justice Council's Online Dispute Resolution Advisory Group, which I had the privilege of chairing (we reported in early 2015), and from the reports on the structure of civil courts (in December 2015 and July 2016) by Lord Justice (now Lord) Briggs, who forcefully advocated the development and launch of online resolution. Sir Geoffrey Vos, the current Master of the Rolls (the most senior civil judge in England and Wales) has taken up the cudgels and is now spearheading modernization and change.

To give a flavour, almost 300,000 money claims have been made using Online Civil Money Claims, a system that enables users to make or respond to money claims on the Web. In the same spirit, are two further services—Damages Claims Online, for making and notifying damages claims; and Possession Claims Online, for helping users to repossess property when owed money for rent or a mortgage, and the tenant or mortgage holder will not pay. The rules governing these and future services will be overseen by the new Online Procedure Rules Committee, established by the recent Judicial Review and Courts Act 2022.

ODR

A field of research and activity that overlaps with online courts is that of 'online dispute resolution', usually known as 'ODR'. It emerged in the 1990s as a branch of ADR (alternative dispute resolution). At that time, ODR was regarded, essentially, as a form of electronic ADR, with many techniques falling under its umbrella, including e-negotiation, e-mediation, and e-arbitration. Rather than meeting in person to conduct, say, mediations and negotiations, various techniques were developed to help sort out a wide range of disagreements informally and on an online basis—from quarrels amongst citizens to conflicts between individuals and the state, from consumer disputes to problems arising from e-commerce.

By way of illustration, an online mediation, or e-mediation, might be undertaken when face-to-face mediation is logistically difficult, perhaps because of the locations of the parties or when, relative to the size of dispute, it is disproportionately expensive to assemble in one forum. In mediation, the mediator, as a third party, assists parties to negotiate settlements, usually on a private and confidential basis. In the same spirit, in an e-mediation, using a mix of human mediators and often technologies— exchanges by email and in online discussion areas—parties to a dispute can settle their disagreements online without convening in a meeting room.

Amongst ODR pioneers, however, the main aim was not to develop systems to support human negotiators and mediators. Many have been more ambitious. The key aspiration of the developers of more advanced systems is that some of the dispute resolution process is, in one way or another, taken on by the system itself.

An early and much touted example was Cybersettle, an online system launched in 1998. In its first version, Cybersettle was claimed to have handled over 200,000 claims of combined value in excess of $1.6 billion. Most of the cases were personal injury or insurance claims. It used a process, much discussed by ODR specialists, known as 'double-blind bidding'—a claimant and defendant each submitted the highest and lowest settlement figures that were acceptable to them. These amounts were not disclosed but if the two ranges overlapped, a settlement could be achieved, the final figure usually being a split down the middle.

A blend of ODR techniques is used to sort out problems on eBay. About 60 million disputes arise each year amongst eBay users. It is unimaginable that these would all get resolved in conventional courts. Instead, ODR is used—swiftly, efficiently, and generally to good effect.

But even these more ambitious systems are fairly primitive, technologically speaking. In the 2020s, I expect that online courts and ODR services will be telepresence-enabled; that is, they will make extensive use of advanced immersive video and metaverse (see later in this chapter). This will allow judges and mediators more easily to communicate with parties in real time. There will be AI-based diagnostic tools to help parties reason about their cases and predictive tools that will make forecasts about the likely outcomes of cases, based on statistical analysis of the past behaviour of courts (this will be like a new version of the ADR process known as 'early neutral evaluation'). More radical yet, in the spirit of game theory, some systems will make concrete recommendations for negotiated resolutions that are sensible outcomes for both parties. And in the private sector certainly, crowdsourcing techniques will be deployed, which will involve decisions being made not by individual judges but by communities of peers.

Blending

Today, people frequently and reasonably use the expressions 'ODR' and 'online courts' interchangeably. But this can be confusing. It is clear that online courts must be an emanation of the state (assuming that 'court' in this context means a public service). Reference to 'ODR', however, is made in both a wide and a narrow sense. In its wide sense, ODR refers, roughly, to any process of resolving a dispute that is largely conducted across the internet. This broad characterization brings 'online courts' under the umbrella of ODR, that is, when online courts use ODR techniques. The narrower sense of ODR equates it more specifically with electronic ADR ('e-ADR'), that is, the systems are regarded as an alternative to public, state-based court service.

A related policy and practical debate is brewing—whether the resolution of, say, low-value disputes should be resolved beyond the public court system, largely by private sector online service providers. A small crop of private offerings has in fact already sprouted and may well become popular in many jurisdictions before state-based online courts are formally launched. Some claim that the market is likely to provide a better and less costly service than the public sector. Others insist that it is a fundamental duty of the state and fundamental to the rule of law that, even in low-value cases, citizens and business should always have realistic access to binding and enforceable dispute resolution by independent judges. In England and Wales, I and others (including the Master of the Rolls and LawtechUK) are advocating a public/private partnership blended arrangement, under which private sector businesses, charities, and others might set up early-settlement ODR facilities (in the narrow sense above), which act as a kind of 'front-end' to the public court system, offering the equivalent of what I call 'extended' services. These ODR front-ends would be accredited by the state; would be linked to the systems of the court, so that unresolved disputes could pass easily into the court's systems (perhaps fast-tracked through); and settlements reached might be enforceable as court orders.

Emerging here we see a combination of state-based courts and online early settlement services that could be much more accessible and preventative than traditional public dispute resolution. For those cases that do require traditional handling by judges, I anticipate another blend of facilities—not just physical courts, but an array of options, including video hearings, audio hearings, hybrid hearings, asynchronous hearings, pop-up hearings rooms, and even hearings in virtual reality.

Robust challenge to a systemic shift away from physical courts is to be expected and indeed encouraged—in the form of open-minded and informed debate. In the digital age, however, given the shortcomings of our current courts, it is reasonable to imagine that various forms of online court and court-connected ODR will be introduced over time. The most fruitful legitimate debate will be over the *extent* of the replacement of conventional hearings. A major outstanding task here is to develop guidelines on what types of technology-based alternative techniques are best suited to what types of dispute.

Fair Trial?

Video hearings, online courts, and advanced ODR may be seen, however, as threatening everyday conceptions of fair trials. For example, victims of crimes and their families, alongside aggrieved and wronged parties in civil disputes, may feel short-changed by a lack of physical meeting. A technology-enabled resolution may not provide the closure that some regard as a central part of the judicial process. Will litigants, more generally, lose their day in court? Yes and no. If a physical hearing is sought for public vindication, then online disposal, although less costly, may fall short. However, if video hearings, online courts, and advanced ODR deliver a much speedier resolution, quicker even than the reasonable time within which justice requires that a case should be heard, then this may well offset the disappointment of not being vindicated in person. It can also be argued—although I contest this—that technology-based solutions should be confined to preliminary hearings and most final trials should be conducted in the traditional manner.

There may be a related concern—that a hearing or trial should be in a publicly accessible forum, so that any wrongdoer's acts are publicly declared and denounced. ODR proponents argue that online facilities will, in due course, offer greater rather than less transparency, because the workings and findings of the court will be observable online in various ways. Interestingly, this concern could equally be a call for televising or broadcasting hearings and processes, which would render them radically more public. This is already happening—the proceedings of the UK Supreme Court are broadcast live on the Sky News website.

As to the actual fairness of decisions, there is no obvious reason why judges or online mediators should be any less impartial, independent, or just when physically remote from some or all litigants, witnesses, and lawyers. It will of course be crucial, in the pursuit of fairness, that there is no actual difference between the soundness of decisions and findings delivered online and those that flow from conventional hearings. It will also be crucial that an online court service is not regarded as an 'economy' service, reserving 'business class' for those who can afford conventional courts. It is not at all obvious that a service that is less costly, quicker, and more convenient and more intelligible will be the inferior service.

Other important questions abound. What about the reliability and credibility of evidence taken remotely? Will judges, juries, and lawyers be at a disadvantage if they cannot look across the courtroom directly into the eyes of witnesses? Or will close-up, three-dimensional video on large, high-resolution monitors permit improved scrutiny? Should lawyers, in hybrid hearings, be with their clients at the camera-end of proceedings or in the hearing rooms near the judges? If the experience of giving evidence remotely is, as is likely, less intimidating than being in a physical courtroom, will this be conducive to evidence that is more or less convincing or decisions that are more or less authoritative and well founded?

More generally, flowing from the thinking of Judith Resnik and Dennis Curtis—in their magisterial book, *Representing Justice*—what will be the impact on public perceptions of justice if one of its main icons, the courtroom, is displaced? Could well-designed online courts—based on user-centric, design thinking techniques (see Chapter 16)—even become symbolic of a new, more inclusive era for dispute resolution? While video hearings, online courts, and advanced ODR may seem alien or outlandish for policy-makers and opinion formers of today, few of these individuals hail from the digital generation. Future generations, for whom working and socializing online will be second nature, may feel very differently. Indeed, for tomorrow's clients, video hearings, online courts, and ODR together may improve access to justice and offer routes to dispute resolution where none would otherwise be available.

It is too early to answer many of the questions just posed in a conclusive way. No doubt, more empirical research and analysis are needed. But, on the face of it, there are no knockdown objections, no overriding concerns of law or principle, that should call a halt to the ongoing and advanced computerization of courts and dispute resolution.

More generally, critics should be cautious about comparing online courts and ODR with some ideal and yet simply unaffordable conventional court service. As Voltaire would no doubt have counselled, 'the best is the enemy of the proportionate'. The comparison that should be made is with what we actually have today—a system that is too expensive, takes too long, is barely understandable to the non-lawyer, and so excludes many potential litigants with credible claims. Regardless of who is paying, we have to find a way of widening access and reducing unmet legal need at a cost that makes sense relative to the value of any given case. In online courts and ODR generally, we believe we have found such a way.

This is why I predict that online courts and ODR will prove to be a disruptive technology that fundamentally challenges the work of traditional litigators and of judges (see Chapter 6). In the long run, I expect them to become the dominant way to resolve all but the most complex and high-value disputes.

Virtual Reality

A more distant possibility is worthy of note—the conduct of a public court hearing in the metaverse or virtual reality (I use the terms interchangeably for current purposes). In broad terms, the idea here would be that participants would enter an alternative, online world and proceedings would be conducted there. These court users would put on headsets and become immersed in computer-generated worlds, which might closely resemble traditional courtrooms or could alternatively look and feel very different. Proceedings might follow conventional court rules or could be governed by simplified or improved processes. In any event, I am imagining proceedings would be largely synchronous, at least in the early years.

Why might we do this? Because human beings often strive to find new ways to conduct old practices. Think of another profession, the clergy—why was there ever a thriving community of Christians running an 'Anglican Cathedral' on Second Life, a virtual online world, with a weekly Bible study class and daily worship services? Daniel Susskind and I discuss this phenomenon in *The Future of the Professions*. Even if some of us might find this bizarre, significant numbers of users found that Second Life Community to be useful and uplifting. The same might be so for legal users of proceedings in the metaverse.

I am not here strongly advocating court hearings on the metaverse. I am simply suggesting them as a possibility. Perhaps some entrepreneurial start-up or notably innovative court official or radical judge might develop a version of this idea and, after some design and development effort, it could gain traction.

Some philosophical support for all of this can be found. There is a conception of justice attributed to Aristotle that speaks of a blind form of judging, when judges are exposed to the case rather than the parties, and the specific characteristics of the parties and their lawyers are not visible. Advocates of this conception of procedural justice could call for virtual reality-based courts in which parties might appear not as faithful renderings of themselves but as avatars, concealing their real identities. The notion of being represented in the metaverse by some sort of avatar may seem outlandish on first encounter. Would this not conceal the real litigant, placing him or her at a distance from proceedings. And yet, in a different way, this is what lawyers are today—a representation of clients and their interests, putting forward cases in their own words and style, rather than in the voice of those who retain them. Today's lawyers are avatars of a sort, although I concede that when clients give evidence, the judge sees them as they are.

Computer Judge?

Lurking behind many debates about online courts is the deeply contentious question of whether computers might ever replace judges. In the early 1980s, I came emphatically to the conclusion that it was neither possible (technically) nor desirable (as a matter of principle) for computers to take on the work of judges. I am now less sure. My position on this is explained at length in *Online Courts and the Future of Justice*. I still accept that judicial decision-making in hard cases, especially when judges are called upon to handle complex issues of principle, policy, and morality, is well beyond the capabilities of current computer systems. I also still accept, for now at least, that most people would prefer a judge to a machine when it comes to resolving potentially life-changing problems. And I accept yet further that most people, in some more general sense, currently regard it as morally questionable if not plain wrong for machines to make judicial decisions.

However, unlike my younger self, I can now conceive of circumstances in which we might even welcome machines generating fully authoritative and binding court determinations. I put this forward as a provocation rather than a full-scale proposal. My starting point is a problem discussed across these pages—in many countries, low-value civil claims are either stuck in great backlogs or, for a variety of reasons, never reach lawyers and courts at all. In Brazil, for example, the backlog in their court system is around 70 million cases, and in most advanced jurisdictions, lawyers are unaffordable for the great majority of people. Whether characterised as chronic delay or unmet legal need, the reality is that cases in neither category are likely ever to be resolved by judges and lawyers in traditional courtrooms. This problem calls for new thinking and new solutions.

What I have in mind is what I call 'predictions as determinations'. Using machine learning technology, we have systems that are getting better and better at predicting the findings of judges (see Chapters 6 and 22). In principle, these systems could be put into operation under the auspices of the state and, with the agreement of parties, it could be enshrined in the rules of court that, for particular types of disputes, if the system predicts a court finding in favour of the claimant with a probability factor greater than, say, 95 per cent, then that finding becomes the official determination of the court.

In the interests of helping parties move on, it is at least possible that some litigants (especially young litigants, for whom online problem-solving is increasingly popular) might say that such a system—one that generates binding predictions, based on data relating to past decisions of the court—is preferable to the gridlocked traditional court or to no accessible resolution process at all.

Again, this idea is not from the realms of fiction. Some jurisdictions—for instance, Brazil, Italy, and Singapore—are actively exploring this possibility. And do remember, if your instinct is that this is bizarre or even dangerous, a key test is not how predictions as determinations compare with traditional judicial decision-making; it is whether they would be better than nothing at all. I refer again to Voltaire.

Readers who have dipped into machine learning might immediately object, pointing out that these systems are black boxes and as such cannot explain their lines of reasoning and, more, they are based on data that may well be biased. In my book on online courts, I respond to these and other concerns at length. I lean in this context, as often I do, toward the pragmatic rather than the purist way forward. And in that spirit, I ask this—even on a cautious or negative analysis of the dangers and limitations of current machine learning, would it not be better on at least some occasions to dispose of cases by predictive systems rather than leave parties to wait for the human judicial decision that will never come? For some court users, the finality of an issue settled might well be more important than the transparency of the process. And when the underlying data is past judicial decisions or data about these decisions, the bias may be no greater than that which inheres in any event in the doctrine of precedent.

15 | *The Future of Law*, Revisited

This book is not my first attempt to make a 20-year prediction in relation to the law and legal services. I was similarly foolhardy in 1996, when I wrote *The Future of Law*. I recall thinking at the time that 2016 seemed like a very long way off, and yet we are already a long way past that milestone. Before we know it, 2036 will be upon us, even if that year sounds today like some distant planet.

When *The Future of Law* was published, only 35 million people were online. Now there are over 5 billion Internet users. Back in 1996, only a few lawyers in the UK had actually seen the World Wide Web, scarcely any clients had received an email from an external legal adviser, and mobile phones were a rarity. Amazon was but two years old. Google had not yet been invented. Nor had YouTube, Facebook, Twitter, TikTok, or Wikipedia. It is at once exciting and disconcerting to contemplate that the life-changing systems and services that will likely emerge in the next 20 years have not yet been conceived.

The subtitle of *The Future of Law* was 'Facing the Challenges of Information Technology'. It is fair to say that this 300-page call to arms caused some consternation in the British legal world. The book was full of outlandish suggestions, such as my promotion of the greater use of email between clients and their lawyers, and my allegation that the Web would be a powerful tool for legal research. The legal establishment was indignant. What was the matter with the postal system? Had I not heard of law libraries? My grip on reality was called into question.

I stand by the broad arguments from that era. In retrospect, however, some of the detail is laughable at best—references to the 'information superhighway', for example, now seem bizarre.

Shift in Legal Paradigm

The central claim of my book was that there was going to be a shift in legal paradigm (a regrettable choice of term in retrospect because 'paradigm' has since become rather overused). By this I meant that many of our fundamental assumptions about the nature of legal service and the nature of legal process would be challenged and changed by the coming of technology and the Internet. In other words, much that we had always taken for granted in the past, about the way that lawyers work and the way non-lawyers receive legal guidance, would be transformed because of the new systems and services that would be built on the back of technology. My 1996 summary of the paradigm shift is reproduced in Table 15.1.

I may not use precisely the same terminology 25 years on, but I think—if I may mark my own homework—that the direction of travel I identified is largely being taken.

TABLE 15.1 The shift in legal paradigm

Today's legal paradigm	Tomorrow's legal paradigm
Legal Service	*Legal Service*
advisory service	information service
one-to-one	one-to-many
reactive service	proactive service
time-based billing	commodity pricing
restrictive	empowering
defensive	pragmatic
legal focus	business focus
Legal Process	*Legal Process*
legal problem-solving	legal risk management
dispute resolution	dispute pre-emption
publication of law	promulgation of law
a dedicated legal profession	legal specialists and information engineers
print-based	IT-based legal systems

At a high level, the first and the last of the changes noted remain the most important. If I will be forgiven for quoting myself, I summarized the move away from advisory service as follows:

IT will eventually enable and encourage legal service to change from being a form of advisory service to a type of information service . . . The ultimate deliverable will be reusable legal guidance and information service pitched at a level of generality considerably higher than the focused advice which characterizes legal advisory work of today.

And, on the final page, in relation to leaving print behind, I predicted that:

legal practice and the administration of justice will no longer be dominated by print and paper in tomorrow's legal paradigm. Instead, legal systems of the information society will evolve rapidly under the considerable influence of ever more powerful information technologies.

Do remember that in 1996, these principal suggestions were regarded as outrageous if not plain seditious. As for the other changes, the filling of the sandwich—it is worth taking stock of each in turn.

Starting with the projected changes in legal service, the move from one-to-one to one-to-many legal service has manifested itself in two ways: in lawyers' capture and re-use of know-how and precedents, and in online service where the content benefits many different users. In both circumstances, rather than handling legal guidance as though it were disposal, it is instead recycled.

There has been much more said than done about greater proactivity in legal service. Sophisticated clients say that they are in the business of legal risk management rather than legal problem-solving, and the ever-burgeoning field of compliance is premised on avoiding rather than resolving legal problems. I expect proactivity will come to the heart of legal service when suitable technologies emerge to help (such as machine learning methods to isolate problems that can be pinpointed in large bodies of agreements or emails).

In relation to time-based billing, commentators for decades have been predicting its demise. At long last, it does indeed seem that there is a steady and widespread shift to commodity-based pricing, in the shape of fixed fees. Lawyers may prefer to charge for the time they spend, but clients prefer to know how much their legal bills will be. Hourly billing will fade away in the 2020s as the dominant charging mechanism.

As guidance on legal issues has become steadily available on the Web, the law has become less alien for many users. Everyday law is already within the grasp of anyone who can read and is connected. To some extent, this has demystified the law. With the ongoing introduction of online dispute resolution and online courts (Chapter 14), citizens and those who run small businesses will increasingly feel empowered and not restricted by the law— it will be within the grasp of everyone to understand and enforce their entitlements.

Most mainstream lawyers in the early 2020s still qualify and hedge their legal advice, often for fear of being sued by those clients (a minority, in my experience) who regard consultation with external counsel as a form of insurance. In contrast, although some online legal services come with disclaimers, most are indeed offered and used in a less defensive and more pragmatic manner. Users seem to understand that a Web-based service is not a human being and although some lawyers have predicted gleefully that they will be kept busy with clients who have suffered loss from relying on defective legal technology, I can find no evidence of this having happened.

The point about shifting from a legal to a business focus is that many problems, for small and global businesses alike, do not come neatly packaged as 'legal'. Rather, legal questions invariably arise in a broader business context and should not be divorced from that broader setting. I predicted in 1996 that online services would be likely to be multidisciplinary in flavour. This is certainly now true of the websites of many government and trade bodies, where legal guidance is often integrated with, say, accounting and tax advice. But lawyers in law firms still organize their businesses and their advice in conventional legal boundaries. It is significant that clients criticize them for this and that the legal businesses of Big 4 accounting firms seem to be focusing more on business solutions than law school subdivisions of the legal world.

Regarding the changes in legal process, the shift from legal problem-solving to legal risk management anticipates a world in which legal problems are often dissolved before needing to be resolved. This links to the swing from reactive to proactive service. The vision here is of a society in which legal problems are contained, prevented from escalating, by the earlier input of legal insight. Historically, this required human lawyers offering early cautionary advice. We now live in a world in which it is second nature for many people to consult the Web to check their rights or obligations. As online legal guidance improves, so too will our ability to identify and manage legal risks.

A related move is that from dispute resolution to dispute pre-emption, which is becoming more visible by the day in the world of online courts (see Chapters 12 and 14). One driving force behind the proposed reforms that we are beginning to observe around the world is an inclination to move cases away from judicial resolution to earlier, often more amicable and less costly handling.

When *The Future of Law* was going to press, there was great debate about the inaccessibility of the law and outrage that it cost a lot to buy the legislation whose binding content we are all assumed to know. A campaign called 'free the law' played an important part in encouraging the British government to make legislative materials available to Internet users at no cost to them. As I wrote in 1996, 'far more materials will be made readily available and easily accessible'. I was referring both to legislation and case law. Two websites corroborate this claim— <http://www.legislation.gov.uk> (the official online database of primary and secondary legislation in the UK) and <https://case law.nationalarchives.gov.uk> (a new (2022) repository of case reports of court and tribunal decisions in England and Wales). But we have some way to go yet before we can say that we fully promulgate, that is, that we have a universally accessible mechanism in place for letting the general public know when new laws are enacted.

Finally, I predicted in 1996 a change in personnel in law— not the disappearance of a dedicated legal profession but the emergence of new roles, and not least the information engineers (we tend today to call them 'knowledge engineers'— see Chapter 16) who work alongside legal specialists in building online legal guidance systems and document automation services. In truth, conventional lawyers still greatly outnumber the new players, but the swiftly growing numbers of lawtech start-ups and technology-based ABSs, alongside the increasing engagement in law of the high-tech large accounting firms and legal publishers, are all clear signs of a new division of labour.

The Latent Legal Market

Looking more broadly again at the shift in legal paradigm, one related claim I made in 1996 concerned the 'latent legal market'. At the time, this attracted a lot of interest. This was my term for the notion that many people in their social and working lives needed legal help and would have benefited from legal guidance but lacked the resources, or perhaps simply the courage, to secure legal counsel from lawyers. As this chapter suggests, much has changed since then—we now have vast online resources available to people who can indeed obtain practical, punchy legal guidance from thousands of government websites or the many sites of the voluntary legal services sector and also of lawyers who offer online legal help as a form of marketing. The latent legal market has to a significant extent been realized.

In all, then, I accept that the shift in paradigm that I projected 20 years ago has not yet fully come to pass. But I submit that the question now is not *whether* the shift will happen but *when* it will transpire. My current expectation is that the transformation will be complete in the 2020s. On a good day, in relation to the predictions in my book, *The Future of Law*, I now feel I was about ten years out.

PART THREE

Prospects for Young Lawyers

16 | New Jobs for Lawyers

In years to come, I predict that conventional lawyers will not be as prominent in society as today. Clients will not be inclined to pay expensive legal advisers for work that can be undertaken by less expert people, supported by capable systems and standard processes. This prediction does not signal the end of lawyers entirely but it does point to a need for fewer traditional lawyers. At the same time, when systems and processes play a more central role in law, this opens up the possibility of important new forms of legal service, and of exciting new jobs for those lawyers who are sufficiently flexible, open-minded, and entrepreneurial to adapt to changing market conditions.

The Expert Trusted Adviser

Two kinds of traditional lawyer will still be in play for the foreseeable future. When work cannot be standardized or computerized, and bespoke service is unavoidable, clients will still call upon their 'expert trusted advisers'. These are intelligent, creative, innovative lawyers who can fashion and articulate new solutions and strategies for clients who have complex or high-value legal challenges (the expert element). These are also lawyers who can communicate their guidance not just with integrity and in strict confidence but in a highly tailored, customized, and personalized way (the trusted component). Many lawyers will say that this is precisely what they do today. They will tell you that all their work already requires expert and trusted handling. Many clients think otherwise and say that lawyers overstate these claims of high craftsmanship. In the end, though, those who handcraft while their competitors are able to source their services in alternative, reliable, and less costly ways will quickly fail to compete.

The Enhanced Practitioner

There will also be a need for the 'enhanced practitioner', a skilled, knowledgeable but not deeply expert lawyer, who will not be asked to deliver a bespoke service but, enhanced by modern techniques of standardization and computerization, will work further to the right-hand side of the evolutionary path that I describe in Chapter 4. The enhanced practitioner will often act as a legal assistant to the expert trusted adviser for those tasks that require a lawyer but not necessarily a costly specialist. Again, however, the market will only have appetite for these kinds of assistant or associate when their legal experience is genuinely needed.

Tomorrow's Lawyers

Although the long-term prospects for most conventional lawyers are much more limited than in the past, I urge young lawyers not to be demotivated or downhearted, because there will be, I believe, a promising range of new opportunities and new careers for people trained in the law. I summarize 15 of these in Table 16.1. I am confident there will be others (see, for example, the more extended discussion in *The Future of the Professions*). For now, these are the jobs that flow quite clearly from the arguments and claims of this book.

When giving talks to law firms and in-house legal departments, I often present a version of Table 16.1 and ask the audience, 'who are these people?' I then face a sea of quizzical expressions. These are the people, I go on to say, who will design the systems that will replace our traditional ways of working. These are the

TABLE 16.1 New jobs for lawyers

legal design thinker
legal knowledge engineer
legal no-coder
legal technologist
legal hybrid
legal process analyst
legal project manager
legal data scientist
legal data visualizer
R&D worker
digital security guard
ODR practitioner
moderator
legal management consultant
legal risk manager

people who will build the systems that will, in the future, solve the problems to which conventional lawyers are currently our only feasible answer. These therefore are the people, I continue, whose systems will solve many of the problems of tomorrow's legal clients. These people, I conclude, are tomorrow's lawyers. At this juncture, the facial expressions morph to incredulity and even indignance. I go on to explain that I prefer to define lawyers as people whose work is devoted to helping legal clients with their legal difficulties, rather than in the more limited terms of their original training and subsequent career path.

If looking at this list unnerves current lawyers, then I am doing my job. This is a wake-up call. Today's lawyers will themselves have to take on some of these roles if they want to remain relevant and in demand. The challenge here is neatly captured in the title of a bestselling book on leadership—*What Got You Here Won't Get You There*. Precisely. Although its author, Marshall Goldsmith, was not focusing on the changing face of the professions, his title sets the agenda for qualified lawyers who are in practice today. In the long run, you will need to change to survive.

Let me now introduce you in more detail to some of tomorrow's lawyers. One day, you may be one or more of the following.

Legal Design Thinker

As more and more legal applications and solutions are built for direct use by people with little legal knowledge, it is vital that legal technology rids itself of its 'designed by lawyers for lawyers' mentality. We need instead to draw from the increasingly popular and successful field of 'design thinking'—techniques, methods, processes, and technologies that can help identify and satisfy the particular needs and wants of the users of systems and services. We need legal design thinkers who will take us way beyond the development of systems that are simply 'user-friendly', although user-friendliness will remain of great importance. Rather, they will inform the actual content and substance of these systems.

For example, in building the next generation of court systems, legal design thinkers will lead us to develop systems that can guide users through complex areas of law; replace the great volumes of court procedure with much leaner rules, many of which will be hidden from users; break down the process into non-forbidding, manageable chunks; and reach beyond the use of text to interact with users, taking advantage of animations, cartoons, videos, flowcharts, and other visual guides to legal processes.

We also need design thinkers to help ensure that the needs of disabled and disadvantaged users are built into the original design of our systems rather than contrived as an afterthought.

Legal Knowledge Engineer

When legal service comes to be standardized and computerized, talented lawyers will be required in great numbers to organize and model huge quantities of complex legal materials and processes. The law will need to be analysed, distilled, and then captured as standard working practices and embodied in computer systems. The result of this might be, for example, an online legal service, or it could be that the law is seamlessly embedded in some broader system or process (see Chapter 6).

Developing legal standards and procedures, and organizing and representing legal knowledge in computer systems, is irreducibly a job of legal research and legal analysis. More than this, it is often more intellectually demanding than traditional legal work, largely because it is more taxing to create a system that can solve many problems than to find an answer to a specific issue. It is plainly wrong to imagine, as many conventional lawyers do, that the development of standards and systems is a task that can be handed over in its entirety to junior lawyers, professional support staff, or even systems analysts. If a modern legal business intends to compete on the strength of its first-rate systems, then it must have first-rate lawyers engaged in building them. These lawyers will be legal knowledge engineers.

Legal No-coder

Some (but not all) legal knowledge engineers will also be no-coders. The idea of no-coding in law can be traced to the 1980s, when people developing rule-based expert systems (see Chapter 22) were greatly assisted by the availability of tools known as 'shells'. Similar systems were available for document automation. If you were able to think logically and reduce the law or legal processes to decision trees or flowcharts then the shells did the rest of the work—they converted your tree or chart into an interactive diagnostic or drafting system. Basically, you did not need to be a programmer to build a serviceable system (although arguably, the process was in fact a form of high-level structured programming).

In the same spirit, a new wave of no-code and low-code tools are emerging in law, enabling the development of systems and applications without involving legions of software engineers. The legal no-coder will be the lawyer who can develop his or her own systems. They will be most useful in creating pilots, demonstrators, and early versions, because generally a fully operational, user-affable system will also require road-testing, tightening, and documenting by a mainstream programmer.

Legal Technologist

The practice of law and the administration of justice have become massively dependent on technology. When legal service becomes impractical or unimaginable without systems, it is vital to have experienced and skilled individuals who can bridge the gap between law and technology. Until recently, two groups have populated the world of legal technology. The first has been made up of mainstream technologists who have found their way into legal environments and have done their best to understand the mysterious ways of lawyers, courts, and clients. The other camp has been occupied by lawyers with a fascination for computers— some were hobbyists, while others attained a more profound appreciation of the world of technology. But neither group, by and large, was populated by professional legal technologists, individuals trained and experienced both in the practice of law and in the profession of systems engineering and technology management. While the technologists and hobbyists worked well enough when technology was largely at the periphery of the delivery of services to clients, we now need a new breed of able and credentialed legal technologists to help take the legal profession fully into the twenty-first century. No longer will it suffice to have mere interpreters, who explain technology matters to lawyers and legal matters to technologists. We require a new cadre of self-sufficient legal technologists whose impact on modern society will be profound—they will lay the foundations upon which legal service is built and the channels through which non-lawyers can access the law.

Legal Hybrid

Lawyers of the future will need to diversify to stay in business. If it is accepted that traditional service will become less common, then I expect lawyers to extend their capabilities by becoming increasingly multidisciplinary. Many lawyers already assert that they are insightful in neighbouring disciplines, and already act, for example, as strategists, management consultants, business advisers, market experts, deal brokers, organizational psychologists, and the rest. Often, with a little probing, it transpires that this experience has been picked up through a brief course or by dipping into an introductory textbook. Although some lawyers like to think otherwise, it is not generally possible to take on a new discipline in a couple of days. Lawyers are highly intelligent human beings (by and large) who are undoubtedly capable of broadening their domains of expertise and becoming first-rate hybrids. But if commercial lawyers want to be strategy consultants, if corporate lawyers aspire to be deal brokers, and if family lawyers wish to be psychologists—and I strongly support this diversification—then this must be backed up by comprehensive and rigorous training that they undertake willingly. The best legal hybrids in years to come will be formidably schooled and unarguably expert in new related disciplines and, in turn, will be able to add considerable value to the legal services they offer to clients.

Legal Process Analyst

I have spoken rather glibly in this book, especially in Chapter 5, about decomposing deals and disputes into their constituent tasks and sourcing these tasks through a multitude of providers. However, the job of analysing a piece of legal work, subdividing the assignment into meaningful and manageable chunks, and identifying the most appropriate supplier of services for each, is itself a task that requires deep legal insight and experience. This will not be an occupation for regular business or systems analysts. It is a job for what I call the 'legal process analyst'. This individual will often be employed within an in-house legal department, for it is not unreasonable for organizations to expect their internal lawyers to be expert at identifying the most efficient and effective way of handling their legal work. Alternatively, legal process analysis could be a service offered by law firms or other third party providers such as accounting firms or legal process outsourcers. Today, there are very few legal process analysts but they are already in demand. Most of the major law firms and in-house legal departments with which I work are very clear that they would readily engage the services of individuals who could undertake reliable, insightful, rigorous, and informed analysis of their central legal processes.

Legal Project Manager

Once the work of the legal process analyst is done, the deal or dispute that has been decomposed and prepared for multi-sourcing will not look after itself. To ensure the success of multi-sourcing, the legal market will require what I call the 'legal project manager'. When the legal process analyst has completed the specification (decomposition and proposed multi-sourcing) it is the job of the legal project manager to allocate work to a selection of appropriate providers, to ensure they complete their decomposed work packages on time and to budget, to control the quality of the various packages, to oversee and supervise the output and delivery, and to pull the various work packages together into one seamless service for the client. This is similar in many ways to the role of the production manager in a manufacturing environment.

The discipline of legal project management should, in my view, be built upon theory and experience from related management disciplines, such as logistics and supply-chain management. No doubt, the legal sector will come to develop its own sophisticated tools and techniques, such as 'legal supply-chain management' and 'legal logistics', which will be core subjects in future courses on legal project management.

Legal Data Scientist

With the growing significance in law of machine learning and predictive analytics, there will be a corresponding need for data experts who are masters of the tools and techniques that are required to capture, analyse, and manipulate great quantities of information. The 'legal data scientist' will seek to identify correlations, trends, patterns, and insights both in legal resources and in non-legal materials. They will also develop systems that can make all manner of legal predictions (see Chapter 6). Again, they will be interdisciplinary specialists, with knowledge not only of relevant systems but also of the law and legal service. A strong background in mathematics, programming, or natural sciences will help here.

Legal Data Visualizer

If a picture is indeed worth a thousand words (see Chapter 13), it is surely time to reduce large bodies of legal documents and complex legal language to more digestible visual forms, such as charts, animations, slides, diagrams, maps, graphs, and so forth. This is work for the legal data visualizer—to bring unprecedented clarity to legal materials. Again, this is an interdisciplinary role, requiring creative talent in producing the visuals as well as legal knowledge to understand the materials. And the stakes are high. The language of the boardroom is no longer text in long documents. It is PowerPoint. Senior executives want to grasp the key issues by absorbing a one-page infographic or placemat, rather than wading through mountains of documents.

R&D Worker

As noted in Chapter 20, the great changes we are witnessing in the legal world mean that the successful providers of tomorrow, like those in today's consumer electronics and pharmaceutical sectors, will need to have research and development (R&D) capabilities if they want to stay competitive. The design and development of new services and solutions will be at the heart of the commercial success of legal businesses in the 2020s and beyond. Lawyers have for long been aware that they have to keep up to date with changes in the law and practice. The 'R&D worker' will have a different focus—on developing new capabilities, techniques, and technologies to deliver legal services in the many different ways anticipated in this book. The efforts of the R&D worker will often be far more exploratory than the daily endeavours of practising lawyers. Dead ends and false starts will not, or should not, be regarded as failures in the R&D function. Indeed absence of failure will suggest a shortage of adventure and flair. The R&D worker will be a very different beast from the typical lawyer of today.

Digital Security Guard

As the dependency on technology and data increases in legal organizations, it will be imperative to have specialists who will ensure their systems and data are reliable and secure, their data privacy is protected, their data is cleansed of bias, their algorithms are subjected to regular audit, and their systems are defended from cyber-attack. These are the basic responsibilities of digital security guards. They are wide-ranging and profound— downtime or data leaks could kill a legal business. Remarkably, many law firms and courts seem to be oblivious to the dangers of neglecting to engage these custodians.

ODR Practitioner

With the advent of online courts and advanced online dispute resolution (ODR—see Chapter 14) as common mechanisms for the settling of disagreements, there will be call for practitioners in this emerging world. These specialists will advise clients on how best to use online facilities and will be experts in resolving disputes conducted in electronic environments. It is early days for services such as e-negotiation and e-mediation, but I have little doubt that imaginative lawyers will, over time, become superior users of these systems and will devise ingenious new techniques that will bestow advantage on those they guide. Litigators need not appear in courtrooms or in video hearings to add value. But they will need to build a new set of skills and methods which position their clients who are involved in online courts and ODR in a demonstrably better position than if they use these systems on their own. New careers will also open up for e-negotiators and e-mediators, those individuals whose intervention and adjudication will be required in the ODR process. Here, as elsewhere in this chapter, the competences that are called for will extend beyond expertise in black-letter law.

Moderator

With the emergence of legal open-sourcing and online legal communities (see Chapter 6), where the recipients of legal guidance convene to exchange their experiences, there will be a growing need for moderators. The problem here is that these facilities and their accumulated insights will be relied upon by lay users. This is not the same as taking advice from and being dependent on traditional lawyers. In contrast, these communities of legal experience are populated by non-experts and so the help and guidance gathered there may be wrong, misleading, unclear, and more. It is important therefore that there is some kind of oversight and quality control of much of the content. The solution is to make moderators available— knowledgeable individuals with relevant experience of the law (and perhaps diplomacy). A law firm, for example, that chooses to set up an online resource through which its clients can share what has worked for them would be well advised to have a moderator, acting as an editor who identifies and corrects errors, inconsistencies, and confusions.

Legal Management Consultant

Many in-house legal departments face a wide variety of management challenges (including, for instance, strategy formulation, restructuring, team building, know-how development, and the introduction of technology). And yet most General Counsel and in-house lawyers have little experience of management issues and so often seek external help. Today, some law firms offer guidance on various management issues but they generally do so in an ad hoc and reactive fashion. Less frequently, professional management consultants are brought in. Given the considerable experience that many law firms have of management issues that arise in their own legal businesses, it is often suggested that these firms might set up their own full-time consulting practices to advise in-house legal departments. Some have done precisely this, and are enjoying success and receiving plaudits. And there is persuasive precedent elsewhere for such a move: in the world of the 'Big 4' accounting firms, their consulting practices originally grew out of their audit businesses; while, more recently, tax management consultancy practices have been built on the back of the work of traditional tax firms.

While the market for this kind of legal management consultancy is still in its youth, it is likely to grow steadily, not just as a way of law firms adding value to their conventional services but as a service line in its own right. Likely services include strategy consulting (on issues such as long-term planning, alternative sourcing, operating models and organizational structure, value-chain analysis of in-house departments, and legal needs assessment) and operational or management consulting (e.g., on recruitment, selection of law firms, panel management, financial control, internal communications, and document management).

Additionally, some legal management consulting providers will offer legal process analysis services. I do not believe that these services can easily or intuitively be provided by lawyers on an informal basis. Instead, legal management consulting will emerge as a distinct discipline.

The Legal Risk Manager

My final category of lawyer for the future is perhaps the most urgently needed and longest overdue. As noted in Chapter 9, most General Counsel consider their primary responsibility to be that of legal risk management. This comes through overwhelmingly in the research I have undertaken in the last decade within the in-house legal community. General Counsel, like the boards to which they report, have a strong preference for avoiding legal problems rather than resolving them. They prefer, as said before, a fence at the top of the cliff to an ambulance at the bottom. What is striking, however, is that there is scarcely a law firm in the world that has acknowledged this need and developed a sophisticated range of processes, methodologies, techniques, or systems to help their clients identify, assess, quantify, hedge, monitor, and control the plethora of risks that confront them. I expect this to change, and the agents of change will be the professional legal risk managers. Whereas conventional legal service is reactive in nature, by which I mean that lawyers (in firms and in-house) spend most of their time responding to questions put to them by their clients, legal risk managers will be proactive (see Chapter 15 for more on this distinction). Their focus will be on anticipating the needs of those they advise, on containing and pre-empting legal problems. Their preoccupation will not be with specific deals and disputes but with potential pitfalls and threats to the business. Legal risk managers will undertake jobs such as legal risk reviews, litigation readiness assessments, compliance audits, and analysis of contractual commitments. Talented legal risk managers will be lawyers with insight into the discipline of risk management and will draw also on techniques from strategy consulting and, increasingly, on emerging techniques for document analysis, such as machine learning (see Chapter 6). This will not be a side-show for the legal profession. It will fundamentally alter the way that clients administer their legal affairs. And the best legal risk managers may, in due course, be eligible for the much wider corporate role of chief risk officer.

Twist to the Tale

Taken together, the 15 new jobs I have identified for tomorrow's lawyers will provide a rich and exciting new set of career opportunities for those who aspire to work in the law for many decades yet. I accept that these jobs are not those that law students generally have in mind when embarking upon law school. But they will be intellectually stimulating and socially significant occupations nonetheless. I know that some lawyers, when they hear of these new jobs, tend to regard them as less prestigious and worthy than traditional consultative service (many craftsmen no doubt felt similarly when their work was industrialized). I respond by saying that those who are already involved in these new roles do consider them rewarding and challenging. Many find themselves contributing in a different but still meaningful way to the higher ideal of an improved justice system.

There is a twist to this tale, however. My guidance to aspiring lawyers who are keen to work in one or more of the fifteen jobs is that it still makes sense, in the first instance, to qualify as a conventional lawyer. This may not be necessary but I think it desirable, not simply because it will for many years yet be useful to enjoy the status of being, say, a qualified solicitor or barrister, but because exposure to and understanding of traditional legal service should provide a valuable foundation upon which to build any new career in law. I am not suggesting that it will not be possible for a law graduate to become a first-rate legal knowledge engineer or legal project manager without having practised as a lawyer. But I do think it would be helpful to have the experience of mainstream legal work.

Lawyers in training should be proactive, however; always on the lookout for experiences that prepare them for tomorrow. I particularly recommend being seconded to clients, working in a variety of countries, and keeping close to the firm's technological developments.

All of that said, I know how difficult it is for many law graduates, in 2022, to secure training contracts and pupillages. This leads, then, to another question. Whether as a conventional lawyer or in one of the new legal jobs, who will employ you in the future?

17 | Who Will Employ Tomorrow's Lawyers?

Sceptics may regard my list of projected law jobs in the last chapter as rather fanciful, not least because they cannot imagine today's law firms or in-house legal departments creating and offering the new roles that I describe.

The reality, however, is that these jobs are already being offered and undertaken in some advanced firms and departments. But the bigger mistake here is to think that these new jobs will always be injected into old legal businesses. This is not how events are likely to unfold. It is more likely that many of these new roles will be offered by a new range of employers who will work in quite different types of legal business. This will be enabled by liberalization (see Chapter 1) and be driven by growing acceptance that many legal tasks no longer require the direct involvement of traditional lawyers. It is not that law firms cannot or will not create the new jobs, but that to do so will often disrupt their conventional businesses. It will simply be easier for new careers and occupations to be fashioned by those businesses, as summarized in Table 17.1, that are currently able to design their future legal services strategy on a blank sheet of paper.

TABLE 17.1 Tomorrow's employers

global accounting firms
major legal publishers
legal know-how providers
legal process outsourcers
big-brand businesses
legal leasing agencies
law companies
online legal service providers
legal management consultancies
lawtech companies

Global Accounting Firms

Most young and aspiring lawyers will not remember the foray of the big accounting firms into the legal market in the early 2000s. The most ambitious of these was Andersen Legal, the legal network of Arthur Andersen, then one of the world's largest accounting and tax practices. At its peak, Andersen Legal had a presence in 30 countries and a total number of lawyers of 2,500. At the time, it was the ninth largest law firm in the world (by revenue). It was growing rapidly, both in size and reputation. Its brand was strong and its multidisciplinary objectives appealed to many clients. It offered a dynamic and exciting working environment, and in my view (biased perhaps, because I was one of their advisers) looked set to redefine the legal market.

Remarkably, Andersen Legal died. Contrary to the common view, however, it did not fail because of some fundamental defect of strategy or business model. Rather, Andersen Legal's demise was a direct consequence of the collapse in 2001 of a giant US corporation called Enron and, in turn, the remarkable implosion of the legal network's parent organization, Arthur Andersen (who were external auditors while Enron perpetrated various accounting frauds). The Andersen debacle did not establish that accounting firms could not provide legal services, but there did follow new bodies of regulations that prohibited these firms from providing other professional services to their audit clients. And this was a notable deterrent, especially in the US, for accounting firms that were contemplating the continuation of their legal services. However, there was not, and is not, a blanket and definitive global prohibition.

In any event, over the past few years, the 'Big 4' (Deloitte, KPMG, PwC, and EY) have begun an emphatic return to the legal world. In fact they never really left and most of them now enjoy many hundreds of millions of pounds worth of annual fee income from their legal services. Liberalization is often said to be a catalyst or justification for their return to the law, and indeed all four have been authorized in England and Wales as 'alternative business structures' (see Chapter 1). However, the more likely attraction is that the legal market, as this book stresses, is of immense value (the value of the global legal market is now not far shy of $1 trillion) and yet is one in a state of flux. The accounting firms see this as a time of great opportunity— they believe they are well equipped to help meet clients' more-for-less challenge and to support their ongoing efforts to manage their legal risks more effectively.

The Big 4's strategies for re-entering the legal market are not yet clear. In the medium term at least, it is unlikely that they will seek to compete directly for the bespoke work of the largest law firms. Instead, they will more likely focus on services and products that leading law firms cannot provide or cannot do so easily. On this view, the Big 4 will be the most formidable alternative legal providers, leading the way in such areas as managed legal services, legal process outsourcing, legal risk management, legal knowledge engineering, legal management consulting, and legal technology.

Whatever competitive route they prefer, these formidable and hugely resourced employers will provide a wide range of career opportunities for tomorrow's lawyers. Already, to give a sense of scale, PwC has more than 3,500 lawyers, working in over 100 different countries and regions.

Major Legal Publishers

Two of the largest legal businesses in the world are Thomson Reuters and RELX (previously, Reed Elsevier). These commercial giants have evolved from the production of conventional print-based publications to the provision of very large and popular legal databases—such as, respectively, Westlaw and LexisNexis (best known amongst lawyers as collections of primary source materials such as legislation and case law). But they have also diversified over the years and are clearly ambitious and acquisitive in the fields of legal technology, legal knowledge engineering, and online legal services. These businesses employ legions of lawyers and armies of software engineers. They are trusted suppliers to the legal profession, and they too see opportunity in the tumultuous legal world. They are commercially ambitious, high-tech, and experienced in carving out new market spaces. They will, unquestionably, provide homes for many law graduates in the future. I think it unlikely that they will deliver conventional consultative advisory legal services but they will offer many of the other jobs noted in the previous chapter. Thomson Reuters is a compelling case study. To deliver their knowledge and research services, they employ huge numbers of qualified lawyers; they own and run a legal process outsourcing company; they bought and now license a leading document automation system; and they have great multitudes of technologists. Accordingly, they can offer a wide range of careers in tomorrow's legal sector.

Legal Know-How Providers

Another class of potential employer for tomorrow's lawyers are legal know-how providers. Nimbler and more entrepreneurial than the major legal publishers, this category was best typified in early years by the English-based legal business originally known as PLC. The company grew steadily from its establishment in the early 1990s, employing many hundreds of lawyers, and operating extensively in England and the US, and was bought in 2013 by Thomson Reuters (as just described). While Practical Law (as the business is now known) is no longer independent and RELX now offers an analogous service known as LexisPSL, I expect to see the emergence of many more know-how providers. We may even see some management buy-outs by the know-how functions within law firms.

Know-how businesses can provide a range of services to law firms and in-house lawyers, including legal research and updates, market intelligence, the provision of standard documents and practice notes, know-how, checklists and flowcharts, playbooks, alongside some conventional law library services. The proposition to the market is that law firms and in-house departments that retain these providers do not themselves then need to maintain their own libraries, information and research services, or professional support lawyers. The know-how providers may not offer legal advice directly but they are engaged in many of the jobs outlined in Chapter 16. And it is easy to see, strategically, how they might progress to providing fuller ranges of legal knowledge engineering, legal process analysis, and legal project management services.

Another way of looking at such businesses—in the language of Chapter 5—is that they are alternative legal providers to which law firms and in-house lawyers will increasingly allocate decomposed tasks. And, as decomposition and multi-sourcing increase, so too will the commercial success and the number of employees of these legal know-how providers.

Legal Process Outsourcers

Perhaps the highest profile of the new, emerging, and alternative legal providers has been the legal process outsourcers (LPOs), businesses that undertake routine and repetitive tasks such as document review in litigation and basic contract drafting. Typically, these businesses—for example, Integreon and Consilio—have established bases in countries in which labour costs are low, such as India. However, LPOs also tend to have significant capabilities in the main jurisdictions in which their clients (law firms and in-house legal departments) are themselves located.

These third party outsourcers are ambitious, entrepreneurial concerns that have often grown rapidly from start-ups, and are unlikely to limit the scope of their services to their current boundaries. Thus, we can expect LPOs to undertake increasingly challenging (and not just the most routine) work, supported by ever more sophisticated processes and systems. And, as liberalization allows, some will doubtless position themselves to offer services that used to be the sole province of law firms. In this way, LPOs may transition into providers of vertically integrated managed services, as discussed below.

LPOs are growing steadily, if not as rapidly as some overzealous commentators have projected. As suggested in Chapter 11, it is in the second stage of the evolution of the legal sector that I anticipate they will peak in their current form. But the best of them too will no doubt adapt and evolve in response to market conditions.

LPOs will be interesting and rewarding employers for tomorrow's lawyers, especially those who are keen on legal knowledge engineering, legal process analysis, legal project management, and compliance process outsourcing.

Big-Brand Businesses

It is only a matter of time, in my view, before huge global businesses move into the law. About a decade ago, for private clients (individuals as opposed to companies), pundits were betting that High (Main) Street banks and supermarkets would soon be heavily involved in the delivery of everyday legal services. Their arguments were attractive. These retail outlets would be more convenient and less forbidding than traditional law firms. More than this, however, as banks and other High Street businesses entered the legal market, it was hoped that they would bring standardization and computerization of routine work, especially high-volume, low-value work. These retail businesses were expected to be direct competitors to traditional law firms, especially small firms.

While we can find examples of this retail development, I expect that major insurance companies, financial advisers, and accounting software suppliers are more likely to move more substantially into law, offering legal services that sit nicely adjacent to their traditional offerings. Less obviously, I also anticipate that big tech companies will in due course enter the fray. I find the idea, for example, of Amazon or Google or Meta offering everyday legal solutions as entirely conceivable, and exciting as well.

While the intuition of many lawyers is to regard these new providers and competitors as removing the need for lawyers, the reality is that the legal services provided by big-brand businesses will need to be developed and often delivered by people of considerable legal experience. Although alternative business structures and others may well be funded and managed by non-lawyers, they will still employ lawyers, old and young. Here, as elsewhere, we should not anticipate that lawyers will no longer be needed but recognize that tomorrow's lawyers may be engaged by quite different businesses.

Legal Leasing Agencies

Another rapidly emerging home for lawyers are legal agencies who make freelance lawyers available to clients. I liken this to leasing lawyers and well-known, long-standing providers of this service are Axiom and LOD (Lawyers on Demand). These businesses offer an alternative career path for lawyers who do not wish to be employed within conventional law firms or in-house departments. For lawyers who want the flexibility to work perhaps six months of the year, such as those with young children, they provide a home of sorts. They have built up a large pool of temporary lawyers who are prepared to work on a contract and project basis. The attraction for the client is that freelance lawyers can be placed within their organizations to meet urgent demands but will tend to cost about half of their equivalent in conventional law firms.

A few legal practices, as I note in Chapter 5, have set up similar businesses that work alongside the traditional firms. Although providing these lawyers in parallel and at lower cost may be seen to cannibalize the firm's traditional work, as I always say of cannibalization—if it is going to happen, you should want to be one of the first to the feast. Thus, entrepreneurial law firms will see opportunities to make the legal experience of lawyers available in new and imaginative ways. For qualified lawyers who want to live a more flexible life, these legal agencies will become increasingly important employers.

It is likely that tomorrow's lawyers (as defined in the previous chapter and engaged in the many roles outlined there) will also be made available on an on-demand basis. Talented legal knowledge engineers or legal data scientists, for example, may find it more flexible, lucrative, and fun not to be fully employed in one organization.

Law Companies

While legal process outsourcers generally focus on undertaking routine, repetitive, administrative, and process-based work and steer clear of substantive legal service, a more ambitious type of alternative provider has emerged. Often called 'law companies', they offer a more vertically integrated service (combining law, process, technology, and consulting) than the LPOs.

High-profile examples here are Elevate and Riverview Law (acquired by EY in 2018). The owners of these and similar legal businesses have started their firms afresh, and jettisoned the old business models. The law companies do not seek to replicate the pyramidic profit structure of law firms, or to bill by the hour, or to work from expensive city buildings. Instead, they keep their overheads very low, they encourage home working (and did so before the pandemic), they have flexible resourcing models, they use technology and knowledge management imaginatively, they outsource their back-office functions, they employ paralegals, and all of this enables them to charge clients less and yet remain profitable. These new-look legal businesses may not offer conventional career paths to partnership, and they may not be as profitable as young lawyers have come to expect of the top firms. But they will offer exciting, vibrant, and entrepreneurial environments in which many young lawyers will thrive in the future. These companies will be open to the overtures of young lawyers who come with ideas of how legal services might be revolutionized. It may even be that they will also be receptive to innovative ways of providing training contracts.

Conversations with law companies tend to be different from interactions with traditional practices. They are less hidebound by past practices, more flexible, and more respectful, I expect, of ideas that emanate from lawyers who are young or young in heart.

Online Legal Service Providers

For young lawyers who are keen to pursue careers as legal knowledge engineers, a natural home may well be a provider of online legal services. Whether providing online advice, online production of documents, or online dispute resolution, these are businesses that analyse and pre-package the law and legal service, allowing clients (both consumers and businesses) to tap into legal insight and service without direct consultation with human lawyers. Deep legal expertise is required, however, in developing these systems and services, and many law graduates and young lawyers in the future will find employment in these businesses.

Although a small number of online legal service providers have achieved remarkable success (LegalZoom springs immediately to mind), it is early days for most of them. But it is hard to imagine, in a digital society, that some will not gain considerable traction. The career opportunities here will vary enormously, from large-scale businesses that will seek, for example, to transform the production of legal documentation in complex deals, through to charitable organizations which will strive to increase access to justice in the ways discussed in Chapter 12.

Legal Management Consultancies

Some traditional consulting firms and some dedicated legal consulting practices will also provide employment for tomorrow's lawyers. These businesses will specialize, for example, in legal process analysis, legal project management, and legal risk management, as well as in advising on the optimum way of managing and organizing a sustainable internal legal function. These specialties may not look like the work that many young and aspiring lawyers have in mind when starting their study of law but they will be central to the legal market and to the interests of clients, nonetheless.

In particular, the demand for legal process analysts and legal project managers will be considerable, so that young lawyers who have taken courses and profess experience in these fields may find themselves more employable than those who boast knowledge only of black-letter law.

Lawyers often speak disparagingly of management consultants. It is true that many advisers who claim to be consultants are less able or experienced than they assert. But there is also a very respectable and impressive body of management consultants whose methods and techniques, I hazard, will be those that many successful legal businesses embrace in years to come.

Lawtech Companies

If client-oriented lawtech companies had existed when I finished my doctorate in law and artificial intelligence (AI) in the mid-1980s, I would have gravitated towards one of them with little hesitation. As noted in Chapter 10, there are now 3,000 to 4,000 lawtech start-ups around the world, many of whom would be well-suited to energetic legal graduates with some experience or interest in the jobs laid out in the previous chapter. Taking employment in these businesses is generally higher risk than working in a law firm—no doubt many will fail or struggle. But a few will fly and it would be exhilarating to be part of one that succeeds.

Many of the start-ups are still at an early stage and may (or may not) seem too precarious a place for employment, but there are also a growing number of more mature independent lawtech businesses—such as Ironclad, Neota Logic, and Litera —that are longer established or well-funded and are forging ahead with exciting new products and services. These are the kind of business that will disrupt the legal marketplace. If you are keen to disrupt and push the boundaries, as a young lawyer, then these lawtech companies may well be your natural home.

Your First Job in Law

If current trends continue, law students may find it increasingly difficult to secure traineeships or employment as young lawyers in conventional law firms. As I say in the previous chapter, if you are graduating from law school, I am still inclined to suggest that you try to obtain work in a law firm or an in-house legal department, so that you can complete your training and qualify as a lawyer. However, if you fail to do so—and this is one of my central and, I hope, positive messages—there will be many other legal businesses, as introduced in this chapter, that may well be interested in engaging you. Alternatively, if you are already qualified and are exploring your options beyond law firms, then exciting new businesses and new options are emerging.

18 | Training Lawyers for What?

I still feel deep sympathy for a law student in a leading US law school who approached me in early 2016, after I had spoken to his class. My lecture had been on the future of legal services and I had covered much of the ground laid out in this book. The student said to me, 'I am almost half a million dollars in student debt and you seem to be saying that my law school is not teaching me the right stuff'. Thankfully, his indebtedness as a US law graduate is unusually high. That said, the average debt of a US law school graduate in 2020, as recorded on the website of the American Bar Association (ABA), was $145,000, which is still a considerable burden.

Most students anticipate that their earnings in law will be sufficient to meet their loans. There are two problems here though. The first is that many law students who might otherwise have been inclined towards, for example, lower-paid public-interest legal work are instead nudged by anxiousness to pay off their debt swiftly towards more lucrative conventional employment. In turn, the so-called third sector generally loses in its war for talent and, more generally, debt leads some graduates along unwanted career paths. The second difficulty is that not all legal graduates secure jobs of any suitable sort.

On this latter point, law schools around the world have for many years been criticized for accepting far more law students than are employable in law firms and other legal organizations. In the US, the point was made very starkly in 2012 by Brian Tamanaha in his book, *Failing Law Schools.* He pointed to government statistics that suggested there would only be 25,000 new openings for young lawyers each year until 2018 while the law schools were annually producing around 45,000 graduates. These figures proved to be pessimistic, although it would be open to Tamanaha to argue that his work might have influenced more responsible numbers of students being admitted to law schools. In any event, today there are approximately 10,000 fewer law graduates in the US than a decade ago. While graduates of top tier law schools stand an excellent chance of securing positions, those who come from lower-ranked schools have a much rougher ride; from some, less than 50 per cent have law jobs on graduating. This overproduction of law graduates is a disorder that can be observed in many other advanced jurisdictions.

For law students who take out enormous loans to undertake their legal studies and do not find work for which they are qualified, it is understandable that they might feel disillusioned. Several years ago, some even tried to raise court actions against law schools for the refund of their tuition fees plus damages, arguing that there was an ongoing fraud in the legal education industry that threatened to leave a generation of law students in dire financial straits. However, students cannot with any credibility advance such arguments today because the issue has been widely ventilated and aspiring lawyers entering law schools know, or at least ought to know, the broad trends.

A related issue is my focus in this chapter—with the appropriateness of what law schools actually teach. My interest is in whether educational bodies are adequately preparing law students for tomorrow's legal marketplace and so maximizing their chances of finding good work. In the US, the ABA Commission on the Future of Legal Education (2017–2019), was clear on the 'crisis', as it put it. The Commission argued that the US system of legal education 'is preparing the next generation of legal professionals for yesterday rather than for tomorrow'. I hope I had some influence on this claim. I was the only non-American on the Commission and had been banging a similar drum for years, not least in earlier versions of this book. Whether the US legal profession will pay heed is not yet clear.

Closer to my home, I am afraid the Legal Education and Training Review in England and Wales, our most thoroughgoing assessment for 30 years, in its report of 2013, largely failed to articulate the education and training needs for the legal industry as it is likely to be. Instead, it provided a very detailed model, in my view, for optimizing the training of yesterday's lawyers. I was also a consultant to this review but did not succeed in convincing the main researchers of the principal messages of this chapter.

Some Assumptions and Concerns

This is not the place for a detailed analysis and evaluation of educational theory and law. However, a number of assumptions and concerns underlie my views on the current and future training of lawyers, and I think it best to lay these out explicitly.

I assume, first, that the law can be an academic discipline worth pursuing for its own sake. There is a contrast, of course, between studying law at university in, say, England as opposed to the US because in the former, this is usually pursued as a first degree, whereas in the latter, the learning of the law is generally undertaken in the form of graduate study. Accordingly, students in the US are more likely, when studying law, to be committed to the legal profession as a career.

When law is studied at an undergraduate level, it can be profoundly stimulating in its own right—the law is one of humanity's most remarkable and sophisticated constructs, a comprehensive system of knowledge that provides a framework for human order and behaviour. To study the substantive rules themselves can be interesting, but to be immersed in subjects such as jurisprudence (broadly, the philosophy of law) and civil law (Roman law) can also be immensely rewarding as self-contained intellectual pursuits. To agree that the academic study of law can be fulfilling in and of itself is emphatically not to suggest that law degrees should be devoted solely to matters of theory. Nor is it to concede that there is no place in a law degree for exposure to legal practice, or insight into the vocation of law, or the acquisition of some of the key skills of the practitioner.

My second assumption is that the training afforded by the pursuit of a law degree can provide useful skills and experience whether or not a career in law is later taken up. Historically, many students have gravitated towards accounting when seeking a broad training to serve them well for a career in the commercial world. But legal training can also be extremely valuable—not just because law graduates have a grasp of a large body of rules and regulations but for the intellectual rigour, the clarity of analysis, the precision with language, the facility for critical thought, the capacity for intensive research, and the confidence in public speaking that a good degree in law should build and provide.

At the same time, third, I am concerned that legal education, delivered in university and in professional exams and sufficient to lead to qualification, is less demanding than that required for many other respectable professions. I look at medicine, architecture, veterinary science, and generally see longer and more arduous courses of study. It is not that studying law at university and taking professional exams is a soft option, but it is certainly quicker and, at least arguably, easier to qualify as a lawyer than to gain entry into many of our other great professions.

My final issue is that the academic and practising branches of the legal profession are insufficiently dovetailed. I look with much admiration, and a little envy, at the great teaching hospitals in London, for example. There, under one roof, a professor of medicine will treat patients, train young doctors, and undertake research, often in the one day. In continental Europe there is a stronger tradition of university law professors also being in legal practice. But in England and to a large extent in the US and Canada, legal practitioners and legal scholars operate in different worlds. In some jurisdictions there is also an unhealthy disrespect in both directions: practising lawyers regard legal academics as ivory-towered theorists divorced from the real world, while legal scholars look upon the daily practice of law as mere business advice rather than serious and substantive black-letter enterprise.

In summary, if you are contemplating the study of law, I can assure you of a stimulating experience that will equip you well for life (assuming it is well taught). However, law schools can be criticized, and often are by practitioners, for not preparing young students adequately for the practice of law.

What Are We Training Young Lawyers to Become?

My own critique of law schools so far has not touched on a far more fundamental concern. It is worrying enough that many practising lawyers question law graduates' preparedness for working in today's law firms. But if graduates are not well equipped for legal practice as currently delivered, they are staggeringly ill-prepared for the legal world of the next decade or two, as anticipated in the earlier parts of this book.

It therefore must be asked: what are we training large numbers of young lawyers to become? This is one of the most fundamental questions of the book. Are we schooling aspiring lawyers to become traditional one-to-one, solo, bespoke, face-to-face, consultative advisers who specialize in the black-letter law of individual jurisdictions and who charge by the hour? Or are we preparing the next generation of lawyers to be more flexible, team-based, technologically sophisticated, commercially astute, hybrid professionals, who are able to transcend legal and professional boundaries, and speak the language of the boardroom and the marketplace? My profound concern is that the emphasis in law schools and professional training is overwhelmingly on the former, with little regard for the latter. Indeed, a more profound concern still is that many legal educators and policymakers do not even know there is a second option. My fear, in short, is that we are training young lawyers to become twentieth-century lawyers and not twenty-first-century lawyers.

To look at this issue in another way, we are focusing, in the training of our lawyers, in the language of Chapter 16, on incubating a new generation of expert trusted advisers and enhanced practitioners but ignoring their likely future careers as legal knowledge engineers, legal technologists, legal process analysts, legal project managers, legal risk managers, and the rest.

It is vital, of course, that we continue to equip young lawyers with the wherewithal to function as first-rate expert trusted advisers and in-house practitioners, but if curricula do not change, it will be neglectful of students and their clients of the future if we do not widen our training to encompass these other new roles.

In many law schools, in terms of content, the law is taught much as it was in the 1970s by professors who have little insight into or interest in the changing legal marketplace. Too often, scant attention is paid to phenomena such as globalization, commoditization, technology, business management, risk assessment, decomposing, and alternative sourcing. And so, I stress again—if many law graduates in the UK are not well prepared for legal work today, they are wholly ill-equipped for tomorrow.

Should we, therefore, extend the remit of law schools and colleges to include other disciplines such as risk management, project management, and legal knowledge management? Is there a place for the future in the busy law curriculum?

The Place for the Future in Legal Education

I am not for a second suggesting that we should jettison core legal subjects such as contract, constitutional law, and tort. Still less am I advocating that we no longer teach students about legal method—how to think like a lawyer, how to marshal and organize a complex set of facts, how to conduct legal research, how to reason with the law (deductively, inductively, analogically), how to interpret legislation and case law, and more. But we do need to think, across the lifecycle of the training of young lawyers, how we can more adequately prepare them for legal practice in the coming decades.

It may be argued that the time and place to train law students in the new disciplines that I identify is not law school but in post-graduate courses, such as the legal practice courses and Bar professional training courses in England and Wales. And there may be a stronger argument still for more intensive training in these emerging fields in the law while undertaking a training contract, pupillage, or some kind of indenture or apprenticeship. But it is clear to me that law schools cannot ignore future practice. Indeed, in my view, it is a dereliction of their duty to do so. What kind of defence can be launched in favour of turning a blind eye, say, to online courts and document automation?

Accordingly, I suggest that we should provide all law students in law schools (and indeed at all stages of their education) first of all with *compulsory* formal immersion into the study of current and future trends in legal services (whether as a stand-alone course or as a significant segment of a course), and second, with the *option* to learn some key twenty-first-century legal skills that will support future law jobs. I do not consider this unduly onerous for law schools, and so I think law students can reasonably demand this of those whom they are paying for the provision of sufficient and appropriate legal education. There is mounting evidence of the need for a legal profession that extends itself beyond traditional service to fields such as legal risk management and legal project management. My case is this: at all stages in the provision of legal education, students should be exposed to systematic teaching about the future of law and legal services, and they should also have the choice and opportunity to be trained in these new skills and disciplines.

Involving practitioners in the delivery of these optional courses would be good, both to give students insight into evolving experience in the market and to encourage cross-fertilization between the academic and practising branches of the profession. I suspect that these courses would be very well attended.

I also have an urgent request of law professors the world over: to take an active interest in the future of legal service; to undertake research (perhaps socio-legal work) into trends in the profession; to expose students to likely futures; and to resist being (in the words of the Dean of a Canadian law school) at 'the cutting edge of tradition'. Law faculties around the world should not just teach a course here or there. Instead they ought to follow the lead of pioneers and innovators in the US, such as the University of Miami School of Law (whose LawWithoutWalls is an exciting, part-virtual, international initiative that aims to change the way that law is taught and practised), Harvard Law School (whose programme on the legal profession is an admirable blend of research, teaching, and collaboration between academics and practitioners), Georgetown Law (with their 'Iron Tech Lawyer Invitational' competition), Chicago-Kent College of Law (offering a J.D. Certificate Program in Legal Innovation + Technology), Michigan State University (which has a Center for Law, Technology & Innovation), and Stanford University with two relevant capabilities (CODEX, a centre for legal informatics, and the Deborah L. Rhode Center on the Legal Profession).

I believe that Europe was first past the post in institutionalizing the field of technology and law in a university law school. I have in mind the Norwegian Research Center for Computers and Law, founded in 1970. We have not, however, followed through on that early promise. With some notable exceptions—for example, Bucerius Law School in Germany—I find most European law schools a little sleepy in their handling of the disruptions that are the focus of this book. In the UK, more specifically, it is with a heavy heart that I write of how few law schools can boast of sustained initiatives devoted either to legal technology or the study of the future of legal services. I am particularly impressed with initiatives at Swansea University, the University of Oxford, Ulster University, Manchester University, and London South Bank University. But the conception of legal practice more widely assumed by the British legal academy is out of step with the prevailing views within the practising branch of the profession on the future of legal service. I fear, if left to their own devices, many law schools will change at a glacial pace. The answer here, I suspect, will be for law firms and in-house legal departments to work more closely with the law schools. The most practical answer is for every law school to appoint a specialist the field, who can evangelize internally and build understanding and support.

Finally, the line of thinking explored in this chapter also provides a new angle on an old question: 'If I want to be a lawyer, should I pursue a law degree as an undergraduate?' I offer no definitive response here but I can see, because tomorrow's legal service will increasingly draw on other fields, that there are stronger arguments now than in the past for studying other disciplines—such as management, computer science, and systems analysis—prior to embarking on a legal career.

19 | Replacing the Old Training Ground

When I speak at conferences, I am invariably asked about the ways in which young lawyers will learn their trade in the future. The concern here is that, according to my hypothesis, a great deal of routine and repetitive work that used to be handled by young lawyers will soon be sourced in different ways, whether by legal process outsourcing, by paralegals, by technology, or the rest. If the basic legal work, upon which young lawyers used to cut their teeth, is to be undertaken by others, how does a young lawyer take the early steps towards becoming an expert?

A Training Problem

This is an important but not fatal challenge to those who advocate alternative sourcing. It is not fatal, in part because this is an obstacle with which very few clients have sympathy. In essence, what we have here is a training problem—alternative sourcing requires law firms to rethink the way they train their lawyers. And most clients, if given the choice, will choose lower-cost legal service from a law firm that has to overhaul its training over a high price service from a firm that seems intransigently wedded to training methods of the past.

The root of the problem here is that because of the so-called war for talent, many firms pay very large salaries to attract the most gifted graduates. However, no matter how talented these trainees and aspiring young lawyers might be, their value for the first couple of years in law offices lies more in their potential than in the actual services they can deliver to clients. The unspoken truth is that the current set-up generally results in clients paying for the training of law firms' aspiring young lawyers. The clients are charged at fairly high hourly rates for the work of these trainees, even though much of the work is process-based and the young lawyers are learning on the job. They may be quick learners but they have insufficient experience or expertise to justify the rates at which they are being charged. Today, in contrast, in changing and more austere times, when clients are demanding more legal service at less cost, they are much less tolerant of paying for the time of budding lawyers who are learning their trade by working on their clients' deals and disputes.

Some years ago, when pondering these issues, I decided to interview some trainee solicitors to determine their take on this conundrum. I spoke to aspiring lawyers who were reviewing endless boxes of documents in preparation for litigation and others who were undertaking major due diligence exercises in support of large deals. I asked them how they would learn their trade if the work they were currently doing was to be, say, outsourced to India. Uniformly they responded that it took only a few hours to learn how to work through piles of documents, and not several months. To express this point more provocatively, we should not confuse training with exploitation. It is disingenuous to suggest that young lawyers are asked to undertake routine legal work largely as a way of them learning their trade. Rather, this delegation has been one mainstay in supporting the pyramidic model of profitability that has enjoyed such unchallenged success until recently.

In any event, it is not at all obvious that aspiring lawyers become expert lawyers by spending months on what is largely administrative work. There is greater evidence that young lawyers learn their trade by working closely with, and observing, legal experts in action.

That said, I do recognize that some work that might be alternatively sourced provides a useful training ground. But it is far from clear how law firms might respond to clients' growing distaste for being charged for the training of their supplier's young people. One likely option, although challenging for many firms, would be no longer to charge clients for trainees and young lawyers unless and until they genuinely bring value. This would directly reduce the profit of those firms that rely on the pyramidic structure. Two possible consequences might follow for young lawyers. The first is that, with the exception of the supremely talented, young lawyers might come to get paid less in their early years of working in law firms than in the past. The second, and more likely result, is that law firms will recruit young and aspiring lawyers in smaller numbers. This does not necessarily mean that all young lawyers will be less employable, because there will be new jobs for lawyers and new employers to work with (see Chapters 16 and 17). But for law graduates, this does indeed appear to be threatening.

Re-Thinking Legal Training

What is the alternative to training lawyers (who aspire to being traditional experts) by putting them to work on relatively straightforward and routine legal matters?

If law firms are genuinely committed to training, I suggest that it should be founded on three basic building blocks in the future. The first is a reversion to some variant of the apprenticeship model. Once young lawyers have their paper qualifications, research and experience suggest that working closely alongside experienced lawyers is a powerful and stimulating way of learning how to move from the law in books to the law in action. If young lawyers are able to share a room with a seasoned practitioner or, as has worked well for years in large accounting firms (and now, falteringly in law firms), to work in open-plan areas with experienced professionals, they will observe and learn, at first hand, effective methods for communicating with and serving clients. If they instead spend most of their time only with other young lawyers and with large boxes of documents, they are less likely to witness and absorb best practice. My support for an in-person apprenticeship presents a challenge to those who are committed to remote working. My approach does not require full-time attendance in the office but certainly assumes some physical presence. At the same time, we should be thinking creatively about techniques for remote apprenticeship—using technology to enable some of the apprenticeship experience to be engendered virtually.

Second, it may be that, when significant bodies of work are sourced from beyond the firm, young lawyers may nevertheless, in parallel, undertake samples of this work, partly to learn their trade and partly perhaps as a way of quality controlling the work done by external providers. In contrast with the past, law firms would need themselves to bear the cost of this work.

Finally, young lawyers should benefit from existing and emerging techniques of e-learning, which in their most advanced forms can be tremendously powerful. This extends beyond online lectures (which themselves can be useful) to online simulated legal practice and virtual legal learning environments. The legal profession's adoption of technology should extend beyond office automation and client service into the way in which we educate and train our young lawyers. Let me say a little more about this.

e-Learning and Simulated Legal Practice

Most senior lawyers and judges of today were legally educated before the birth of the personal computer. Lectures, tutorials, and countless hours in dusty law libraries were the order of the day. And much of the time was spent memorizing vast lists of case names and sections of statutes, alongside potted summaries of their significance. None of this should be uncritically inherited by tomorrow's law schools and by the colleges that offer conversion and legal practice courses.

Take traditional one-hour lectures as an example. There are now compelling arguments, based on cost and on the reality of what actually goes on in a lecture theatre, for reserving conventional live lectures for the relatively rare occasions when wonderful speakers are scheduled and expected to deliver memorable social or educational experiences. Here is the truth about most conventional law lectures in the UK—they are often not given by gifted (or even trained) orators. Some professors mumble and ramble, others simply read from their notes, while only a very few are inspirational. No wonder the attendance is low. And this phenomenon persists across the country in countless law schools. It is wasteful and insufficiently edifying. There is no good reason for not replacing the ramblers and the dictators by online lectures, which can be presented by wonderful and inspirational speakers (selected from across the land) who make webcasts of their lectures available. Anyone who has used TED (<http://www.ted.com>) will know how powerful an online talk can be. During the pandemic, of course, some students have benefited from attending outstanding online lectures. However, I suspect that the majority have not—a great university lecture is rare; a great online rendition is even rarer (few professors have many years of experience of communicating by video link).

Alternatively, live lectures of a sort can be given as real-time webinars, which allow students to participate and debate throughout. Online resources can be made available to help students prepare for these sessions. I became convinced of the potential of a hybrid approach in 2009 when I was asked to undertake a five-year review of e-learning at the then College of Law in England (now the University of Law). I found that electronic tutorials and online supervision had changed the learning experience of law students on the college's legal practice course. More than 400 'i-tutorials' had been developed—online, 'head and shoulders' webcasts by legal experts with slides on the side. Students found these mini-lectures highly convenient in that they could be stopped, started, and replayed, on hand-held devices as well as on laptops. The college went a step further and developed one-to-one supervision by tutors, conducted virtually rather than face to face. This created what I called an 'electronic Oxbridge'—many of the strengths of the traditional tutorial system were secured (the pressure, the inspiration, and the attention of a personal expert tutor) but achieved in a practical, affordable way. This approach, by necessity, became commonplace during the pandemic, even if not always seamlessly executed. It is time now to reflect on what works well online and what does not, and to industrialize this first generation of online legal learning and blend it with more traditional, physical sessions.

Online lectures, i-tutorials, and virtual supervision, however, are only part of the future for legal education. The pioneering work of Paul Maharg moved us beyond this first generation into a world of simulation-based training and transactional learning, as described in his book, *Transforming Legal Education*. He pioneered these techniques in the post-graduate Scottish Diploma of Legal Practice taught at Strathclyde University for which he designed a fictional town, 'Ardcalloch', in which law students played the role of lawyers in virtual law firms. An early example of a legal metaverse, the students practised law—simulations of actual legal transactions and disputes— with experienced solicitors acting as clients and judges in this virtual environment. All sorts of facilities were made available online: virtual offices, various institutions, professional networks, and with a collection of documentary resources to lend authenticity, including newspaper clippings, photographs, wills, bank books, and advertisements. I have no doubt that students immersed in such a simulated working environment, with practising lawyers participating and supporting, could be exposed to a more profound and memorable learning experience, that greatly outstrips ill-attended lectures and non-participatory tutorials.

These e-learning techniques will only become more powerful—exposure to advocacy, drafting, client meetings, negotiations, document reviews, due diligence exercises, and much more will increasingly be provided online. If we can develop flight simulators for astronauts, we should be able to put together facilities that will be immeasurably more effective in training young lawyers than asking them to review endless piles of documents or memorize lists of cases. And they will go some considerable way to patching the gaps left by alternative sourcing.

20 | Questions to Ask Employers

In this chapter, I change emphasis. If you are a young or aspiring lawyer applying for a new job, I want now to equip you with some questions to pose when, at the end of a gruelling interview, you are faced with the inevitable query: 'Do you have anything you would like to ask us?' I also recommend these questions to you if you are a young lawyer wondering whether your current firm is one to which you should commit yourself in the long run. Note that these questions are very similar to those that I ask of law firm leaders when they engage me as an external strategy consultant. Together, they seek to determine the depth of an organization's insight into the future and its appetite for change.

The questions come with a health warning. There are quite a few of them and, although it is good at a job interview to be engaged, savvy, and interested, I am not suggesting you fire off all the queries at the one sitting. It is generally counter-productive at interviews to appear excessively objectionable or subversive. Also, I am conscious that the job market is so intensely competitive that many readers will be glad to secure any position at all, so that these questions might seem of peripheral concern. However, it is good to be armed with some penetrating observations and it should be helpful and relevant for tomorrow's lawyers to give serious thought to some difficult issues.

Do You Have a Long-Term Strategy?

This simple query can prompt all sorts of physical reactions, from nervous giggles to disparaging grunts. Law firm leaders often respond by claiming that they have not actually written down their strategy in a formal document but that all the partners know what their strategy is. Invariably, this is nonsense. In such firms, most partners will confess in private to having no clue as to the strategy of their business. The leaders themselves are either dissembling or rationalizing. It is not that a strategy document in and of itself has great value but an absence of such a document usually betrays an absence of strategic thinking.

Beware of law firm leaders who say that in the current economic climate their focus must be on the short term. As stressed in Chapter 8, the best leaders keep one eye on short term issues and the other on the long-term strategic health of their organizations. Worry deeply if a senior partner is preoccupied with 'low-hanging fruit' or 'quick wins'. This frequently betrays short-termism which can precede rapid decline.

A quite different response is the production of a fat 300-page report. An external firm of management consultants will often have produced this. This of itself is problematic—the composition of the strategy of a firm, the document which seeks to determine its very future, is too important a task to outsource to another organization. Moreover, it should not take several hundred pages to articulate the strategy for a legal business.

You are unlikely ever to be handed a full strategy document, which will be guarded as though 'top secret'. But you may get a sanitized précis. What you should look for in this is evidence of a firm that has thought deeply about changes in the broader business environment and in the legal market in particular. There should be a sense from the document of the ambition of the firm, of where it hopes to be in, say, five years' time, and what major changes it must effect to get there. It should indicate the markets in which it intends to work and how it will seek to compete in these chosen markets. You should also find an indication of the values that are central to the firm, and of the culture it likes to engender. You should be convinced that the overall sense of strategic direction seems realistic. What to search for here is a relatively small number of major priorities, rather than a litany of piecemeal initiatives.

If a strategy document with such contents does not exist, then this is not a business that is preparing wisely for the future, and not a business, therefore, that is likely to provide a firm foundation for lawyers of tomorrow.

What Will Legal Service Look Like in 2040?

My previous question, about long-term strategy, tests the business's view over perhaps the next five to ten years. This next query, about the way in which legal service might change over the next generation, is looking around 20 years out. In the introduction to this book, I observed that when I was at law school in the early 1980s and was discussing the future with my friends and professors, there seemed to be a shared view that the basic daily work of lawyers would be much the same a quarter of a century hence. In the event, we were not wrong. At that time, there were no obvious and imminent drivers of change, analogous to the more-for-less pressure, liberalization, and technology (the running themes of this book).

Looking at technology in particular, although IBM launched its PC when I was in the penultimate year of my law degree, we did not yet live in a time of massive and foreseeable technological upheaval. In radical contrast, given that we are now witnessing an exponential increase in the uptake and power of technology, it would be remarkably short-sighted for anyone to maintain that legal service in, say, 2040 will be much the same as today. Of course, no one can predict what our world will be like then but in asking a prospective employer a question about the distant future, you should not be looking for a definitive, authoritative response. Indeed, be wary of anyone who is excessively dogmatic in any direction. The kind of firm in which you should want to build your career, if you are convinced by the arguments and predictions of this book, is a firm in which its members express both interest and concern about the years ahead. A dismissive response is a narrow-minded response, and you should be looking, in stark contrast, for a firm that is open-minded and welcoming of debate about possible futures.

Are You Comforted by Other Firms' Lack of Progress?

If change is unavoidable, then bright lawyers in impressive law firms will usually adapt promptly and effectively. They have no choice. If there is a burning platform of some sort, they have no option but to jump off. In the absence of such an imperative, most law firms, even the finest, tend to be driven more by a fear of lagging behind their competition than by a hunger for forging ahead of their rivals. Law firms, in other words, are motivated more by the need to avoid competitive disadvantage than a thirst for the attainment of competitive advantage. This is quite unlike many other sectors, such as consumer electronics, where the driving passion is continuously to outpace and out-think the competition. When meeting with law firm leaders, I find the easiest way to motivate them is to speak of the notable achievements of their closest rivals.

It follows, therefore, that many lawyers do indeed derive great comfort from knowing that other firms, of whom they think highly, have made but modest attempts to rethink the way they work, or to embrace technology, or to take up many of the suggestions in this book.

Accordingly, you should view with great optimism a firm that insists that it is driven not by its competitors but by its clients' needs, that the market will clearly require fundamental changes, and that the conservatism of other firms presents an opportunity for a new market leader to emerge. If these are the messages that you hear from an organization, strive to gain employment with it.

Interestingly, if you meet with alternative providers in the legal marketplace, such as legal process outsourcers, legal publishers, or large accounting firms, then you will detect amongst them a far greater appetite for change, far greater excitement about the future, than that evinced by mainstream law firms in their often rather subdued response to shifting market conditions.

What are Your Preferred Approaches to Alternative Sourcing?

If this question is met with a blank expression or a whiff of incomprehension, it may need a little explaining, as follows: given that clients are increasingly asking their legal advisers to find ways of reducing the costs of routine and repetitive work, which approaches are you finding to be the most promising?

Treat with suspicion those firms that say no more than that they are looking very carefully at this challenge, or that they are currently in discussions with low-cost providers such as legal process outsourcers, freelance lawyers, or the like. Look for evidence of action and activity rather than of reflection and discussion.

Even if firms say they have invested in some facility—perhaps a near-shore centre or subcontracting arrangement—probe a little further to test if this is tokenism or serious commitment. A number of firms do indeed have arrangements in place but these are often no more than gestures that allow the partners to say to clients that they are involved in some way.

Ordinarily, it will be clear if a firm is genuinely committed. You will sense their enthusiasm and they will have tales to tell about what has worked well and what needs further refinement.

What Role Will Technology Play in Law Firms of the Future?

Most senior lawyers are still not entirely comfortable when talking about the changing role of technology in their firms. They will speak articulately enough about the systems they currently use, such as email, word processing, PowerPoint, accounting and billing, as well as their handheld devices to which they seem permanently tethered. Most law firms will have sophisticated technology departments and their dependence on technology is indeed profound. However, the technologies I have in mind are not back-office systems but those that directly affect and support client services. One category, for example, is knowledge systems—the collection of applications (from intranets through know-how databases to internal social networks) which seek to capture and make available a firm's collective experience and expertise ('bottom-right' systems, in the language of Chapter 7); or client relationship systems, those services such as online deal rooms that provide communication and collaboration channels between firms and their clients ('top-left'); or online legal services—systems that, for instance, provide legal guidance and document automation ('top-right').

In the 2020s, we will see a shift of emphasis from back-office systems to technologies that transform, often disruptively (see

Chapter 6), the traditional way that lawyers have worked with their clients.

To gauge the technological sophistication of a firm that you are considering, you should look in the first instance for recognition of the kinds of change just described, and then for evidence of investment in these emerging technologies. One interesting sub-question to pose is: 'What is the formal process by which your firm monitors emerging technologies and evaluates their potential for your various practice areas?' You will find that very few firms have such a process. If you locate one that does, look no further.

Do You Have a Research and Development Capability?

If you are a consumer electronics company, like Apple or Sony, you have not yet invented the products that will be the foundations of your business in five years' time. The position is similar in pharmaceutical companies. This is why these and many other businesses have research and development (R&D) budgets and departments—teams of very bright people (that I envisage wearing white coats and having high brows) who are given freedom to think deeply and creatively and to come up with all sorts of possible offerings for the future. Most of their ideas never see the light of day. The R&D people are encouraged to think the unthinkable, to be bold and daring. And if their inventions are not commercialized, this is not regarded as failure.

There is an analogous challenge here for law firms. If what I say in Parts One and Two of this book comes to pass, then lawyers too, perhaps five or ten years from now, will be offering services that they have not yet conceived. So, how are these firms going to innovate? Who is going to come up with new, market-changing legal services? It is not unreasonable to ask law firms whether they invest in R&D and, if so, in what way. A supplementary question here might be to ask what percentage of their annual fee income is ploughed back into an R&D budget (consumer electronics and pharmaceutical companies spend about 15 to 20 per cent of their turnover on R&D).

Very few firms currently have R&D budgets or departments, and so an encouraging reply would be that they know that this is soon going to be necessary. Firms that are dismissive and hope they can for some years squeeze yet more juice out of the old way of working should be viewed with some distrust as long-term places of employment.

If You Could Design a Law Firm from Scratch, What Would it be Like?

In my consulting work with clients, I have built a formal exercise around this question. I call it 'blank sheet thinking'. I have discovered that most lawyers, when thinking about the long term, tend to be contained and constrained by their current set-up. Their thinking ahead is legacy-based; they are walking backwards into the future. In contrast, in these times of great change, I encourage law firms to be vision-based, to put to one side the way their firms are currently organized and positioned, and take a leap forward to consider where they could and should be in, say, five years' time.

To help them to engage in this vision-based thinking, I ask them to answer the following question: 'If you were given a blank sheet of paper and could design your practice or firm from scratch, what would it look like?' (I provide a series of prompts to support them but they need not detain us here.) Pursuing a similar line of inquiry, try to get a sense from your prospective employers of what would be different if they could wave a magic wand and, in response to current and impending pressures, were able to build their business afresh.

You may find, as I do, that this thought experiment releases lawyers from their focus on current ways of working and reveals some fascinating insights about where law firms might be located, how many people they would employ, how they would source their work, what technology they might deploy, the extent to which they would seek external capital, and much more.

If the upshot of the interviewers' responses to your questions is that their business would look much as the one they are managing today, then I would be deeply sceptical. On the other hand, if this question leads to a series of imaginative and engaging thoughts about different ways of working, then that employer may well be an exciting prospect for you.

Remember that I am not suggesting that you should bombard prospective employers with this last question and all the others outlined in this chapter. Nonetheless, it is impressive to be armed with an astute question or two, and the answers to the queries suggested here could be highly illuminating for your future.

21 | Innovation

Many lawyers and law firms are now claiming to be innovators. This apparent commitment to substantial change is peppered across their websites, press releases, publications, as well as their pitches to clients. And it is fuelled by a veritable industry of innovation awards, lauding those who are said to have pushed boundaries. The reality, for now, is more sobering. If I sound a little cynical in this chapter, this is because I am. I remain in no doubt that the legal world is in need of pervasive change and that this change will indeed come to pass. And, to be clear, I am still excited by many of the advances I anticipate. But much of the current wave of innovation in traditional law firms falls short of the overhaul I envisage and recommend. This is an illustration of what I dismiss in the preface of this book—the suggestion that tomorrow's lawyers have already arrived.

First- and Second-Generation Legal Innovators

From my ongoing research and countless visits, I find that most law firms that say they are innovating are in fact what

Daniel Susskind and I call, in our updated edition of *The Future of the Professions*, first-generation and not second-generation innovators. In broad terms, although first-generation innovators often speak of their projects with great flourish, their preoccupation with change, on analysis, is usually with efficiency gains. The second generation, by contrast, is pursuing more radical change. The difference between the two is summarized in Table 21.1, by reference to ten sets of features which, we have found, are common to all professional firms and not just law firms.

Taking each of the ten features in turn, the innovation activities of the majority of law firms, in the first place, focus more on improving today's working practices and processes than on implementing novel business models. First-generation

TABLE 21.1 First- and second-generation innovation in law

First generation	Second generation
• process improvements	• new business models
• marketing noise	• substantive progress
• automation	• innovation
• pilots	• fully operational systems
• little impact on figures	• significant revenue and profit
• argument-based	• evidence-based
• minority of partners involved	• majority of partners involved
• intellectual grasp	• emotional commitment
• avoiding competitive disadvantage	• seeking competitive advantage
• short-term, tactical	• long-term, strategic

legal innovators tend to work on streamlining and optimizing the conventional one-to-one consultative advisory model of professional service, whereas the second generation are in the more ambitious hunt for sustainable alternatives to selling the time of human lawyers.

The cold truth of first-generation innovation is often obscured by the garrulous promotional activities of law firms that choose to position themselves as transformational and disruptive. This is the land of innovation-by-press-release. In contrast, the second generation broadcast less and get on with execution. They value reality over perception. By analogy with comedians, the first generation assert they are funny, while the second generation tell jokes. There are less polite ways of putting this.

What about legal technology? The first-generation firms focus on automation—they computerize or systematize their traditional ways of practising law. They graft lawtech onto current processes. The second generation take a different approach. They embrace innovation, in the sense I discuss in Chapter 1—they are developing legal products, services, and solutions that were not feasible (or even conceivable) without technology. Also, the first generation tend to deliver prototypes or proofs-of-concept while the second generation deliver polished final systems. Lawyers who are new to lawtech should appreciate that there is a yawning chasm between a pilot and a fully operational system.

Regarding the figures, first-generation innovators, with their demonstrators and marketing noise, seldom have significant impact on financial performance. This contrasts markedly with the second generation whose innovations generate not just reputational payback but creditable turnover and profit. And there is nothing like fee income to whet the appetite of most law firm partners.

In turn, this means that, internally, the second generation is better regarded than the first because they have tangible evidence of the impact of their innovations. They have case studies they can share and these boost confidence and encourage further investment. By contrast, with no practical and positive outcomes to share, the first generation relies more on argument than evidence. Commercially minded lawyers, frankly, are more taken with income than with abstract theories of, say, disruptive innovation.

This in large part explains why first-generation law firms can boast of only a small number of partners who are wholly committed to innovation—these are advocates of radical change for whom argument and theory are enough. The landscape in second-generation firms is very different. Here, where there is plentiful evidence of the commercial benefits of innovation, most partners are seriously enthused. Partners in first-generation firms do of course grasp the arguments in favour of innovation. But they do not feel it in their hearts and bones. They are not emotionally invested, whereas second-generation partners are already beneficiaries and feel they are a part of a different and unfolding future.

When it comes to strategy, first-generation law firms participate in defensive innovation. They concentrate more on not being left behind than on leaping ahead of competitors. They focus on avoiding competitive disadvantage. Contrast the second-generation firms who engage in proactive innovation. They regard innovation as an opportunity rather a threat. Their aim is to secure sustainable competitive advantage. The strategies of the first- and second-generation firms are therefore very distinct, and clients can sense the difference more readily than most lawyers believe.

Innovators of the first generation are short-term operators, tactical thinkers who are consumed with the trappings of change and harvesting low hanging fruit. By contrast, legal innovators of the second generation—rare beasts—are longer-range planners who weave radical change into all corners of their strategic planning processes.

Innovation and Transformation

In summary and in truth, as with lawtech, we are still at the foothills of innovation in the world of law. On inspection, progress in most law firms has been less striking than might appear at a glance from a distance. In my experience, the same goes for 'transformation' in law firms. The majority that I visit (admittedly not a random selection) have some kind of transformation programme up and running. However, I have noticed that most of these transformation initiatives—like innovation initiatives—tend to end up as efficiency projects. They are launched optimistically with lofty aspirations of effecting radical and lasting change, but, after 12 to 18 months, most partners become impatient, they clamour for 'quick wins' and 'low hanging fruit', leadership relents, the transformation plans are watered down, and the focus shifts instead to streamlining and optimizing what is already in place.

It is considerably easier to speak from a podium or write a press notice about transformation than it is to put it into practice.

When confronted with proposals that are truly transformative, most lawyers recoil a little. They prefer a lighter hand on the tiller, and find it more palatable and less threatening to advance iteratively from where they currently are than to envision and advance towards a wholly different end state. This is entirely understandable because no-holds-barred transformation is genuinely disruptive and so often destructive of current business models and services. And it is far from clear whether it is feasible or desirable for successful law firms to self-disrupt. It is easier to sidestep self-destruction. In my travels, I have not yet found any major law firm that has intentionally and pervasively transformed itself and in so doing significantly self-disrupted. Experience from other sectors and industries suggests that the disruption and even destruction is more likely to be wrought by new players such as the lawtech start-ups (or upstarts) than the law firm incumbents.

Innovation and Differentiation

While on the subject of disruption, I think it worth clarifying a confusion over two different types of innovation—'blue ocean strategy' and 'disruptive innovation'. Legal blogs are packed with references to both, frequently dropped into discussion as though they are the same beast. They are not. The former has its origins in an excellent book, *Blue Ocean Strategy*, by Renee Mauborgne and W Chan Kim (first edition in 2004), while the latter finds its best exposition in another first-rate work, *The Innovator's Dilemma* by Clayton Christensen (first published in 1997). The idea of blue ocean strategy is to create and then dominate *new* uncontested markets, that is, to originate, stimulate, and then satisfy entirely new demand. Strictly, the idea of disruptive innovation is about fundamentally transforming existing markets. This, then, is a red ocean strategy—it is about competing in markets that currently exist (the red denoting the blood that flows from a feeding frenzy where there is fierce competition). Blue ocean strategy is, broadly speaking, a diversification strategy (doing new things), while disruptive innovation is a transformation strategy (overhauling existing offerings). Both are innovation strategies.

Whichever strategy is preferred, my first plea is for more action and less rhetoric in the world of legal innovation. My second plea is that we should be more ambitious in our pursuit of innovation. My feelings and views in this context are captured memorably by Theodore Levitt in *Thinking about Management*, where he links the concept of innovation to differentiation:

> Innovation must constantly look for possibilities of product differentiation. In the pursuit of customers, every seller seeks to be in some appealing way different from all other sellers. Optimally, every seller tries to be irresistibly different. The aim is to become so distinctively different as to be, in effect, a monopolist of the offering—to be perceived as being not just the best provider of the particular product, but as being its *only* provider. Ideally, everybody else should be seen as being in a different and lesser league.

I love that. *Irresistibly different.* To be sure, Levitt sets the bar high. But he also compels us to take innovation seriously.

22 Artificial Intelligence and the Long Term

In the long run, the changes that I anticipate for lawyers and the administration of justice will be pervasive, irreversible, and transformational. I am not suggesting that this means the legal sector will be turned on its head over the next three to six months, but I am confident we will see many fundamental shifts as we move through the 2020s.

Looking further ahead, by 2040, to pick a date two decades or so hence that will be mid-career for young lawyers of today, it is neither hyperbolic nor fanciful to expect that the legal profession will have changed beyond recognition. In this final chapter, my purpose is to put this legal revolution in some wider context.

Artificial Intelligence

In thinking about the long term in law, it is hard to ignore the current avalanche of interest in artificial intelligence (AI) for lawyers. Barely a week goes by without a new claim in the press or social media about a robot lawyer or AI-based system that is better than or has even replaced traditional human lawyers.

The majority of major law firms in the UK, for example, have signed up with one AI provider or another and have expressed great hopes for their investments. I am intrigued by these claims because I have a long-standing interest in the field—in the mid-1980s, I completed my doctorate on AI and law at Oxford and have been fascinated by relevant developments ever since.In my view, most of the current claims being made by enthusiasts and pundits about the impact of AI on the law greatly *overstate* its likely *short-term* impact. However, and more importantly, most of the current claims about the *long-term* greatly *understate* the impact of AI on legal service. Will AI transform the work of lawyers over the next couple of years? No, it will not. What about by 2030? I expect it most certainly will. This at least is my expectation, as I begin my fifth decade of working on AI in the law. I have not felt able to say this until now. This analysis suggests that the 2020s will be the decade during which artificial intelligence will take hold in law.

Our machines and systems are becoming increasingly capable, and, over time, they will take on more and more legal tasks that we have historically regarded as the unique preserve of legal practitioners. This seems to be what people have in mind today when they speak of AI in law—systems that undertake various categories of legal work that in the past required thinking, human lawyers. In particular, in the terminology of Chapter 6, recent discussion of AI has focused on document analysis, machine prediction, legal question answering, and, to a lesser extent, document automation. No-one is suggesting that these systems are actually conscious (they are examples, therefore, of what is known as 'weak AI' rather than 'strong AI') but, functionally, they seem to be doing some of the work of lawyers.

When I began work in the field of AI and law, one approach dominated—the knowledge and reasoning processes of legal experts were mined from their heads through an interview process known as 'knowledge elicitation'. That knowledge was then codified in complex decision trees, and dropped into computer systems, around which non-expert users could navigate. We called them rule-based expert systems. They put questions to users and were able to provide legal answers and draft legal documents, often at a higher standard than human experts. In 1988, I co-developed the world's first commercial system (the Latent Damage System) which advised in this way on a corner of the law of limitation. The system reduced research time from hours to minutes and the subject matter expert, Professor Phillip Capper, happily admitted that the final version outperformed him. But these systems were costly to build and maintain. And they held little attraction for law firms because they reduced the time taken to do legal work, which was not appealing in an era dominated by lavish and largely uncontested hourly billing.

Although sceptics say this first wave of AI had little impact, its underlying techniques are still used widely today—for example, in document automation systems around the world, as well as in online legal services offered by law firms. More, the multi-billion-dollar tax compliance industry (personal and corporate tax) is built on this first wave of AI in law.

Importantly, there is now a second wave of AI and its developers reject the early idea that the way to get machines to solve legal problems is for them to copy the best human experts. Three types of system are crucial. First are those that can analyse huge bodies of legal materials. This is the world of 'machine learning' and 'Big Data'. Some systems can already make better predictions than expert lawyers. Drawing on data from more than 100,000 past cases, Lex Machina, for instance, can predict the probability of success in US patent litigation more accurately, it has been claimed, than litigators. A related family of systems (initially 'trained' or 'supervised' by lawyers) can search through massive litigation bundles and identify relevant documents more precisely than junior lawyers and paralegals. Similar techniques are used for due diligence. These systems are identified as 'disruptive' in Chapter 6.

Also disruptive are systems, second, that answer questions and solve problems in an apparently intelligent manner (legal question answering). The latest question answering systems draw on remarkable recent advances in natural language processing (see Chapters 1 and 6). My favourite illustration of a system that could apparently answer questions is IBM's Watson, the system that, as noted in Chapter 1, appeared in 2011 in a live broadcast of an American TV quiz show, and beat the two best-ever human contestants. Think about this—a system that, effectively, answered questions about anything in the world, more rapidly and accurately than any human being. Inspired by early success in the use of Watson in medicine, many law firms and legal suppliers have explored deployment of analogous techniques in law.

Finally, there is the field of affective computing, which is delivering systems that can detect and express emotions. Systems can already differentiate between a fake smile and genuine smile more reliably than humans.

When machines today can make predictions, identify relevant documents, answer questions, and gauge emotions at a higher standard than human beings, it is not just reasonable but vital that we ask whether people or systems will be doing our legal work in decades to come.

And yet, many lawyers still steadfastly insist that their work cannot be replaced by machines. They say that computers cannot think or feel and so cannot exercise judgment or be empathetic. This claim usually rests on what Daniel Susskind and I, in our book, *The Future of the Professions*, call the 'AI fallacy'—the view that the only way to get machines to outperform the best human lawyers will be to copy the way that human lawyers work. The error here is not recognizing that the second wave of AI systems do not replicate human reasoning. We saw this in 1997 when IBM's Deep Blue system beat the world chess champion, Garry Kasparov. It did so not by copying the thought processes of grandmasters but by being able to explore up to 330 million moves per second. So too in law—human lawyers will be outgunned by brute processing power and remarkable algorithms, operating on large bodies of data.

I say again, therefore, as our machines become increasingly capable, they will steadily eat into lawyers' jobs. The best and the brightest human professionals will last the longest—those experts who perform tasks that cannot or should not be replaced by machines. But there will not be enough of these tasks to keep armies of traditional lawyers in employment. This is not an imminent threat to lawyers. In the 2020s at least, as explained in Chapter 16, there will be redeployment and not unemployment for lawyers—lawyers will undertake different work. In career terms during this period, lawyers should plan either to *compete* with machines (look for legal jobs that are likely to favour human capabilities over artificial intelligence) or to *build* the machines (aim to be directly involved in the development and delivery of new legal technologies and systems). In the very long term, it is hard to avoid the conclusion that there will be much less need for conventional lawyers. If this is so, then 'build' is a better career strategy and indeed business strategy than 'compete'. More than this, many young lawyers will be more excited by the prospect of being involved in building the systems that will replace our old ways or working than by a job that entails the perpetuation of our outdated practices.

Power Drills or Holes?

Understandably, this line of argument about AI and replacing some or much of the work of lawyers generates considerable nervousness and indignation when floated before lawyers and law students. I often find it helpful to invite the sceptic to consider one of my favourite business stories, concerning a leading manufacturer of power tools. It is said that this company takes its new executives on a training course when they join up, and at the outset a slide on a large screen is presented for their consideration. The slide is a photograph of a gleaming power drill and the presenters ask the assembled new recruits if this is what the company sells.

The new executives seem slightly surprised by this but together buck up the collective courage to concede that 'yes, this is indeed what the company sells'. With a flourish and some evident satisfaction, the trainers then ask for the next slide, which depicts an image of a hole, neatly drilled in a wall. 'This', they say, 'is actually what our customers want, and it is your job as new executives to find ever more creative, imaginative, and competitive ways of giving our customers what they want.'

There is a great lesson here for lawyers. Most senior legal practitioners, when contemplating the future of their business, tend to be of power drill mentality. They ask themselves, 'What do we do today?' (answer: one-to-one, consultative advisory service, often on an hourly billing basis) and then, 'How can we make this service cheaper, quicker, or in some way better?' Very rarely do they take a step back and ask themselves, by analogy, about the hole in the wall in the legal world. What value, what benefits, do clients really seek when they instruct their lawyers?

For 25 years and more, I have been asking lawyers, 'What is the hole in the wall in the delivery of legal services?' For me, one of the best answers to this question came indirectly from KPMG, one of the world's leading accounting and tax firms. I am not a great fan of mission statements or the like, but on KPMG's website, some years ago, I noticed part of one that I thought was superb: 'we exist to turn our knowledge into value for the benefit of our clients'. I think this is a great way of capturing the value that lawyers bring: lawyers have knowledge, expertise, experience, insight, know-how, and understanding that they can apply in the particular circumstances of their clients' affairs. Lawyers have knowledge and experience that their clients lack.

Notice that KPMG did not say that they exist to provide one-to-one consultative, advisory services on an hourly billing basis. They did not confuse their methods of working with the value they deliver.

Many insights flow from KPMG's rendition of the role of the professional or legal adviser. For lawyers, the most significant for me is a challenge that follows from it—what if we could find new, innovative ways of allowing our clients to tap into our knowledge and expertise? What if we, as lawyers, could make our knowledge and expertise available through a wide range of online legal services, whether for the drafting of documents or for the resolution of disputes or by embedding sound legal and compliance practice in our clients' working processes? If we can find online or embedded methods of enabling access to our experience and the service is thereby less costly, less cumbersome, more convenient, and quicker, then I suggest that clients, oppressed as they are by the more-for-less challenge, would welcome these services with arms flung open.

But surely, it is often objected, clients always want a human lawyer they can trust. My research suggests otherwise. People with legal problems want a solution that is trustworthy—if this can be delivered online in a way that fixes their problems reliably, they will often happily forego human service.

No Change is the Least Likely Future

It is often observed, not especially profoundly, that we cannot predict the future. This seems to give licence to the unimaginative, the short-sighted, and the indolent to discard any foresights as pointless speculation. In contrast, I join others who believe that we can anticipate many (but not all) of the broad contours, if not the specific details, of the world yet to come.

One interesting way to think about the future is to contemplate the sustainability of what we currently have. Given our economic conditions, the shift towards liberalization, the new providers in the marketplace, and the burgeoning, exponential increase in the power and uptake of technology, I find it unimaginable that our current legal institutions and legal profession will remain substantially unchanged over the next decade. Indeed, it seems to me that the least likely future is that little will change in the world of law. And yet, the strategies of most law firms, law schools, and in-house legal departments still assume just that. In truth, for much of the legal market, the current model is not simply unsustainable. It is already broken.

Look at the law and legal service from another vantage point. At the heart of law and legal service is legal information (from raw law such as legislation through to deep expertise held in specialists' heads). Pause now and think about information. We are currently witnessing a change in the information sub-structure of society. This is the term that I introduced in 1996 to refer to the main way in which information is captured, shared, and disseminated. I share the view of anthropologists who have observed that human beings have travelled through four stages of information sub-structure: the age of orality, where communication was dominated by speech; the era of script; then print; and now a world in which communication is increasingly enabled by digital technology. There will no doubt be a fifth, when nanotechnology, robotics, genetics, and technology converge, perhaps in 30 or 40 years' time. In this era of transhumanism, my guess—and I say this with some hesitation because it could easily be stripped from context by critics—is that entire bodies of law and regulation will then be embedded in chips and networks that themselves will be implanted in our working practices or, eventually even, in or remotely accessible to our brains.

For now, we are coming to the end of the transitional phase between the third and fourth stages of development, between a print-based industrial society and an AI-based digital society. The key point here is that the information sub-structure in society determines to a large extent how much law we have, how complex it is, how regularly it changes, and those who are able, responsibly and knowledgeably, to advise upon it. If we examine the manner in which the law has evolved throughout history, we can understand the shifts in terms of changes in information sub-structure. At its core, then, law is information-based. And we are in the middle of an information revolution. It is not making a wild leap to suggest that the law and the work of lawyers will not emerge unscathed.

This thinking led me in 1996, in my book *The Future of Law*, to predict the shift in legal paradigm discussed in Chapter 15, by which I meant that many or most of our fundamental assumptions about legal service and legal process would be challenged and displaced by technology and the Internet. It was a 20-year prediction, and I hazard again that the direction of travel I sketched out has proven to be accurate, although I concede we are running almost ten years behind.

Do We Need a Legal 'Profession'?

The changes I anticipate in this book raise further and deeper questions about the future of professional service. Why is it that we give monopoly rights to certain occupational groups over particular areas of human endeavour? The accountancy profession, medical profession, and the legal profession, for example, are exclusively entitled and permitted, respectively, to conduct statutory audits, to perform surgery, and to engage in advocacy in upper courtrooms. It is as though there is some social contract—in *The Future of the Professions*, we call this the 'grand bargain'—that empowers certain skilled and knowledgeable classes of people to undertake work that it would be foolhardy or dangerous for lay people to attempt on their own. Thus, we have these trusted advisers, who are responsible for keeping their knowledge current and applying that knowledge in a confidential, affordable, and accessible way. We trust these individuals because of their training and experience, their competence and integrity, and their codes of conduct. And they enjoy the reputation and prestige of groups whose experience is respectfully called upon by fellow human beings.

There are, however, several problems with this model. In the first place, in most societies, we struggle to afford the delivery of professional knowledge and experience in a one-to-one, conventional manner. In these stressed economic times, health services, legal services, educational services, and many more are under enormous strain. The old model does not seem to yield readily affordable and accessible service.

The second challenge to the model is that a new channel for the delivery of knowledge and experience has been developed. This is digital technology. As indicated throughout this book, it is possible to allow lay people to tap into the insights and experience of lawyers through, for example, online legal guidance systems, document automation systems, and communities of legal experience.

A third challenge to the profession probes into the heart of a crucial issue—the motivation of those who are in opposition to change. Building on Clay Shirky's quotation at the start of this book, it is leaders as well as institutions within professions who try to preserve the problem to which they are the solution. In more common parlance, turkeys rarely step forward to vote for an early Christmas. There are none so conservative or reactionary as those who benefit from the status quo. It is no doubt this line of thinking that led George Bernard Shaw famously to claim that 'all professions are conspiracies against the laity'.

I put it a little differently. I observe, in law, that there are two distinct camps (and a few in between): the benevolent custodians and the jealous guards. The benevolent custodians are those who, consistent with the conception of professionalism just noted, regard it as their duty to nurture the law and make it affordable and accessible to members of society. They are the interface between lay people and the law and they strive to be user-friendly. In contrast, the jealous guards wish to ring-fence areas of legal practice and make it their exclusive preserve, whether or not the activity genuinely requires the experience of lawyers and with little regard to the impact of this quasi-protectionism on the affordability and availability of legal service. In the US, when lawyers object to online legal services that help citizens and claim the providers are engaged in the unauthorized practice of law, we frequently see this second camp in action. The disingenuity of their claims—that their primary concern is access to justice or safeguarding the interests of their clients—makes me shudder. In truth, for many (but not all), their primary concern is themselves and threats to their income and their self-esteem.

Your Mission

I implore you, tomorrow's lawyers, to take up the mantle of the benevolent custodians; to be honest with yourselves and with society about those areas of legal endeavour that genuinely must be preserved for lawyers in the interests of clients. But you should work in the law in the interests of society and not of lawyers. Where, in all conscience, legal services can responsibly and reliably be offered as technology-enabled self-service for non-lawyers, celebrate access to justice and draw upon your creative and entrepreneurial talents to find other ways that your legal knowledge and experience can bring unique value to your clients.

As I often remind lawyers, the law is no more there to provide a living for lawyers than ill health exists to offer a livelihood for doctors. It is not the purpose of law to keep lawyers in business. The purpose of lawyers is to help to support society's needs of the law.

Alan Kay, a computer scientist from Silicon Valley, makes a different but related point. He once said that 'the best way to predict the future is to invent it'. This is a powerful message for tomorrow's lawyers. The future of legal service is not already out there, in some sense pre-articulated and just waiting to unfold. It is not that I, and other commentators who follow trends in legal services, can see the future where most lawyers cannot. All that I do is lay out a metaphorical buffet—a set of possible courses that lawyers and other legal service providers may or may not choose.

Here is the great excitement for tomorrow's lawyers. As never before, there is an opportunity to be involved in shaping the next generation of legal services. You will still find most senior lawyers to be of little guidance in this quest. Most of your legal elders will tend to be cautious, protective, conservative, if not reactionary. They will resist change and will often want to hang on to their traditional ways of working, even if these are well past their sell-by date.

In truth, you may feel you are on your own. But I urge you to come together, to join a growing movement of people who I say are 'upgrading justice'—exploiting technology in forging new paths for the law, our most important social institution.

FURTHER READING

There is a rapidly growing literature on the future of legal services. In the short bibliography here, I include the publications explicitly mentioned in the main body of the book together with a selection of texts and longer articles and resources that I recommend to readers who want to delve further into the field. I have not included references for all law firms, legal businesses, public reports, and online services that are noted in the book: they can easily be found online.

PUBLICATIONS

American Bar Association Commission on the Future of Legal Education, *Principles for the Future of Legal Education and Licensure in the 21st Century* (Chicago: American Bar Association, 2020).

Armour, J. and Sako, M., 'Lawtech: Levelling the Playing Field in Legal Services?' (21 April 2021). Available at SSRN: <https://ssrn.com/abstract=3831481>.

Baker, S., *Final Jeopardy: Man vs. Machine and the Quest to Know Everything* (New York: Houghton Mifflin Harcourt, 2011).

Beaton, G. and Kaschner, I., *Remaking Law Firms* (Chicago: American Bar Association, 2016).

Benkler, Y., *The Wealth of Networks* (New Haven: Yale University Press, 2006).

Black, N., *Cloud Computing for Lawyers* (Chicago: American Bar Association, 2012).

Bull, C., *The Legal Process Improvement Toolkit* (London: Ark, 2012).

Canadian Bar Association, *Futures: Transforming the Delivery of Legal Services in Canada* (Ottawa: Canadian Bar Association, 2014).

Chambliss, E., Knake, R.N., and Nelson, R.L., 'What We Need to Know About the Future of Legal Services' (2016) 67(2) *South Carolina Law Review* 193.

Chan, K.W. and Mauborgne, R., *Blue Ocean Strategy* (Boston: Harvard Business School Press, 2005).

Christensen, C., *The Innovator's Dilemma* (Boston: Harvard Business School Press, 1997).

Civil Justice Council (Online Dispute Resolution Advisory Group), *Online Dispute Resolution for Low Value Civil Claims* (London: Civil Justice Council, 2015).

Cohen, M.A., '"New Law?": You Ain't Seen Nothing Yet', *Forbes*, 31 May 2022, available at <https://www.forbes.com/sites/markcohen1/2022/05/31/new-law-you-aint-seen-nothin-yet/?sh=7ed4eb4f104e> (accessed 9 August 2022).

Croft, J., 'Why Are Investors Pouring Money into Legal Technology', *Financial Times*, 28 July 2022, available at <https://www.ft.com/content/b6f0796e-0265-40c6-ad4c-a900cd788c39> (accessed 9 August 2022).

Dershowitz, A., *Letters to a Young Lawyer* (New York: Basic Books, 2001).

DeStefano, M., *Legal Upheaval* (Chicago: American Bar Association, 2018).

Dutton, W. and Blank, G., *Cultures of the Internet: The Internet in Britain, Oxford Internet Study 2013* (Oxford: Oxford Internet Institute, 2013).

Faure, T., *Smarter Law* (New York: Thomson Reuters, 2018).

Galbenski, D., *Unbound: How Entrepreneurship is Dramatically Transforming Legal Services Today* (Royal Oak: Lumen Legal, 2009).

Goldsmith, M., *What Got You Here Won't Get You There* (New York: Hachette, 2021).

Grossman, M. and Cormack, G., 'Technology-Assisted Review in E-Discovery Can be More Effective and More Efficient Than Exhaustive Manual Review' (2011) XVII(3) *Richmond Journal of Law and Technology* 1.

Hagan, M., 'Participatory Design for Innovation in Access to Justice' (2019) 148(1) *Daedalus* 120.

Harper, S.J., *The Lawyer Bubble* (New York: Basic Books, 2013).

Haskins, P.A. (ed.), *The Relevant Lawyer* (Chicago: American Bar Association, 2013).

Kafka, F., *The Trial* (Harmondsworth: Penguin, 1983).

Katsh, E. and Rabinovich-Einy, O., *Digital Justice: Technology and the Internet of Conflict* (New York: Oxford University Press, 2017).

Katz, D.M., Bommarito, M.J., and Blackman J., 'Predicting the Behavior of the Supreme Court of the United States: A General Approach' (21 July 2014), available at SSRN: <http://ssrn.com/abstract=2463244>.

Kimbro, S., *Limited Scope Legal Services: Unbundling and the Self-Help Client* (Chicago: American Bar Association, 2012).

Kowalski, M., *Avoiding Extinction: Reimagining Legal Services for the 21st Century* (Chicago: American Bar Association, 2012).

Kowalski, M., *The Great Legal Reformation* (Bloomington: iUniverse, 2017).

Kurzweil, R., *The Singularity is Near* (New York: Viking, 2005).

Jacob, K., Schindler, D., and Strathausen, R., *Liquid Legal* (Berlin: Springer, 2017).

JUSTICE, *What is a Court?* (London: Justice, 2016).

Kohlmeier A. and Klemola M., *The Legal Design Book* (published by authors, 2021), available at <https://legaldesign-book.com/> (accessed 9 August 2022).

Law Society of England and Wales, *The Future of Legal Services* (London: The Law Society, 2016).

Legal Services Board, *Understanding Consumer Needs from Legal Information Sources* (London: LSB, 2012).

Levitt, T., 'Production-Line Approach to Service' (1972) (September–October) *Harvard Business Review* 41.

Levitt, T., *Thinking about Management* (New York: The Free Press, 1991).

Levitt, T., *Marketing Myopia* (Boston: Harvard Business School Publishing Corporation, 2008).

Levy, S., *Legal Project Management* (Seattle: DayPack, 2009).

Lightfoot, C., *Tomorrow's Naked Lawyer* (London: Ark, 2014).

Maharg, P., *Transforming Legal Education* (Aldershot: Ashgate, 2007).

Morgan, T., *The Vanishing American Lawyer* (New York: Oxford University Press, 2010).

Mountain, D.R., 'Disrupting Conventional Law Firm Business Models Using Document Assembly' (2007) 15(2) *International Journal of Law and Information Technology* 170.

Newton, J., *The Client Centered Law Firm* (Vancouver: Blue Check Publishing, 2020).

Paliwala, A. (ed.), *A History of Legal Informatics* (Saragossa: Prensas Universitarias de Zaragoza, 2010).

Parsons, M., *Effective Knowledge Management for Law Firms* (New York: Oxford University Press, 2004).

Paterson, A., *Lawyers and the Public Good* (Cambridge: Cambridge University Press, 2012).

Pink, D., *A Whole New Mind* (London: Cyan, 2005).

Rainey, D., Katsh, E., and Wahab, M. (eds), *Online Dispute Resolution: Theory and Practice* (2nd edn, The Hague: Eleven International, 2021).

Regan, M. and Heenan, P., 'Supply Chains and Porous Boundaries: The Disaggregation of Legal Services' (2010) 78 *Fordham Law Review* 2137.

Resnik, J. and Curtis, D., *Representing Justice* (New Haven: Yale University Press, 2011).

Rhode, D.L., *The Trouble with Lawyers* (New York: Oxford University Press, 2015).

Sako, M., *General Counsel with Power?* (Oxford: Said Business School, 2011), available at <http://www.sbs.ox.ac.uk/>.

Schroeter, J. (ed.), *After Shock* (Chicago: Abundant World Institute, 2020).

Staudt, R.W. and Lauritsen, M. (eds), 'Justice, Lawyering and Legal Education in the Digital Age' (2013) 88(3) *Chicago Kent Law Review* 879.

Susskind, R.E., *Online Courts and the Future of Justice* (Oxford: Oxford University Press, 2019; paperback edn, 2021).

Susskind, R.E., *The End of Lawyers?* (Oxford: Oxford University Press, 2008; paperback edn, 2010).

Susskind, R.E., *The Future of Law* (Oxford: Oxford University Press, 1996; paperback edn, 1998).

Susskind, R.E., *Transforming the Law* (Oxford: Oxford University Press, 2000; paperback edn, 2003).

Susskind, R.E. and Susskind, D.R., *The Future of the Professions* (Oxford: Oxford University Press, 2015; paperback edn, 2022).

Tamanaha, B., *Failing Law Schools* (Chicago: University of Chicago Press, 2012).

Tellman, P., *Building an Outstanding Legal Team* (Woking: *Global Law & Business*, 2017).

Tromans, R., 'Artificial Lawyer', available at <https://www.artificiallaw yer.com/> (accessed 9 August 2022).

Woolf, Lord, 'Access to Justice—Interim Report and Final Report' (Woolf Inquiry Team, June 2005 and July 2006), available at <http://www.justice.gov.uk/>.

INDEX

*For the benefit of digital users, indexed terms that span two pages (e.g., 52–53)
may, on occasion, appear on only one of those pages.*

Tables are indicated by *t* following the page number

275